ASSESSMENT ITEM LISTING FOR ELECTRICITY AND MAGNETISM

HOLT, RINEHART AND WINSTON

A Harcourt Classroom Education Company

Austin • **New York** • **Orlando** • **Atlanta** • **San Francisco** • **Boston** • **Dallas** • **Toronto** • **London**

Copyright © by Holt, Rinehart and Winston

All rights reserved. No part of this publication may be reproduced or transmitted in any form or by any means, electronic or mechanical, including photocopy, recording, or any information storage and retrieval system, without permission in writing from the publisher.

Teachers using HOLT SCIENCE AND TECHNOLOGY may photocopy complete pages in sufficient quantities for classroom use only and not for resale.

Art and Photo Credits

All work, unless otherwise noted, contributed by Holt, Rinehart and Winston.

Front cover: IT Stock International/Index Stock Imagery/Picture Quest; (owl on cover, title page) Kim Taylor/Bruce Coleman, Inc.

Printed in the United States of America

ISBN 0-03-065532-3

1 2 3 4 5 6 082 05 04 03 02 01

CONTENTS

Introduction v
Installation and Startup vi
Getting Started viii
1   Introduction to Electricity 1
2   Electromagnetism 27
3   Electronic Technology 52

Introduction

The *Holt Science and Technology* Test Generator and *Assessment Item Listing*
The *Holt Science and Technology* Test Generator consists of a comprehensive bank of test items and the ExamView® Pro 3.0 software, which enables you to produce your own tests based on the items in the Test Generator and items you create yourself. Both Macintosh® and Windows® versions of the Test Generator are included on the *Holt Science and Technology* One-Stop Planner with Test Generator. Directions on pp. vi–vii of this book explain how to install the program on your computer. This *Assessment Item Listing* is a printout of all the test items in the *Holt Science and Technology* Test Generator.

ExamView Software
ExamView enables you to quickly create printed and on-line tests. You can enter your own questions in a variety of formats, including true/false, multiple choice, completion, problem, short answer, and essay. The program also allows you to customize the content and appearance of the tests you create.

Test Items
The *Holt Science and Technology* Test Generator contains a file of test items for each chapter of the textbook. Each item is correlated to the chapter objectives in the textbook and by difficulty level.

Item Codes
As you browse through this *Assessment Item Listing*, you will see that all test items of the same type appear under an identifying head. Each item is coded to assist you with item selection. Following is an explanation of the codes.

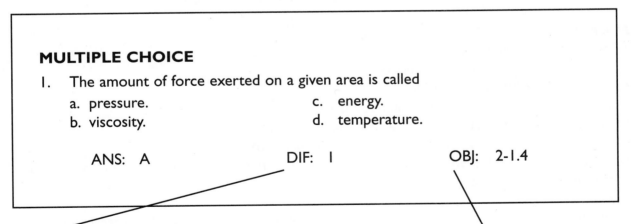

MULTIPLE CHOICE
1. The amount of force exerted on a given area is called
   a. pressure.
   b. viscosity.
   c. energy.
   d. temperature.

   ANS: A   DIF: I   OBJ: 2-1.4

DIF defines the difficulty of the item.
  I requires recall of information.
  II requires analysis and interpretation of known information.
  III requires application of knowledge to new situations.

OBJ lists the chapter number, section number, and objective.
(2-1.4 = Chapter 2, Section 1, Objective 4)

INSTALLATION AND STARTUP

The Test Generator is provided on the One-Stop Planner. The Test Generator includes ExamView and all of the questions for the corresponding textbook. ExamView includes three components: Test Builder, Question Bank Editor, and Test Player. The Test Builder includes options to create, edit, print, and save tests. The Question Bank Editor lets you create or edit question banks. The Test Player is a separate program that your students can use to take on-line* (computerized or LAN-based) tests. Please refer to the ExamView User's Guide on the One-Stop Planner for complete instructions.

Before you can use the Test Generator, you must install ExamView and the test banks on your hard drive. The system requirements, installation instructions, and startup procedures are provided below.

SYSTEM REQUIREMENTS

To use ExamView, your computer must meet or exceed the following hardware requirements:

Windows®
- Pentium processor
- Windows 95®, Windows 98®, Windows 2000® (or a more recent version)
- color monitor (VGA-compatible)
- CD-ROM and/or high-density floppy disk drive
- hard drive with at least 7 MB space available
- 8 MB available memory (16 MB memory recommended)
- an Internet connection (if you wish to access the Internet testing features)*

Macintosh®
- PowerPC processor, 100 MHz
- System 7.5 (or a more recent version)
- color monitor (VGA-compatible)
- CD-ROM and/or high-density floppy disk drive
- hard drive with at least 7 MB space available
- 8 MB available memory (16 MB memory recommended)
- an Internet connection with System 8.6 or a more recent version (if you wish to access the Internet testing features)*

* You can use the Test Player to host tests on your personal or school Web site or local area network (LAN) at no additional charge. The ExamView Web site's Internet test-hosting service must be purchased separately. Visit www.examview.com to learn more.

INSTALLATION

Instructions for installing ExamView from the CD-ROM:

Windows®
Step 1
Turn on your computer.
Step 2
Insert the One-Stop Planner into the CD-ROM drive.
Step 3
Click the Start button on the taskbar, and choose the Run option.
Step 4
In the Open box, type "d:\setup.exe" (substitute the letter for your drive if it is not d:) and click OK.
Step 5
Follow the prompts on the screen to complete the installation process.

Macintosh®
Step 1
Turn on your computer.
Step 2
Insert the One-Stop Planner into the CD-ROM drive. When the CD-ROM icon appears on the desktop, double-click the icon.
Step 3
Double-click the ExamView Pro Installer icon.
Step 4
Follow the prompts on the screen to complete the installation process.

Instructions for installing ExamView from the Main Menu of the One-Stop Planner (Macintosh® or Windows®):

Follow steps 1 and 2 from above.
Step 3
Double-click One-Stop.pdf. (If you do not have Adobe Acrobat® Reader installed on your computer, install it before proceeding by clicking Reader Installer.)
Step 4
To advance to the Main Menu, click anywhere on the title screen.
Step 5
Click the Test Generator button.
Step 6
Click the appropriate Install ExamView button.
Step 7
Follow the prompts on the screen to complete the installation process.

GETTING STARTED

After you complete the installation process, follow these instructions to start ExamView. See the ExamView User's Guide on the One-Stop Planner for further instructions on the program's options for creating a test and editing a question bank.

Startup Instructions

Step 1
Turn on the computer.
Step 2
Windows®: Click the Start button on the taskbar. Highlight the Programs menu, and locate the ExamView Pro Test Generator folder. Select the ExamView Pro option to start the software.
Macintosh®: Locate and open the ExamView Pro folder. Double-click the ExamView Pro icon.
Step 3
The first time you run the software, you will be prompted to enter your name, school/institution name, and city/state. You are now ready to begin using ExamView.
Step 4
Each time you start ExamView, the Startup menu appears. Choose one of the options shown.
Step 5
Use ExamView to create a test or edit questions in a question bank.

Technical Support

If you have any questions about the Test Generator, call the Holt, Rinehart and Winston technical support line at 1-800-323-9239, Monday through Friday, 7:00 A.M. to 6:00 P.M., Central Standard Time. You can contact the Technical Support Center on the Internet at http://www.hrwtechsupport.com or by e-mail at tsc@hrwtechsupport.com.

Short Course N Chapter 1—Introduction to Electricity

TRUE/FALSE

1. If a switch is closed, charges flow freely through the circuit.

   ANS: T   DIF: I   OBJ: 1-4.1

2. The loads in a parallel circuit do not necessarily all have the same amount of current in them.

   ANS: T   DIF: I   OBJ: 1-4.3

3. When a short circuit occurs, resistance increases and current decreases.

   ANS: F   DIF: I   OBJ: 1-4.3

MULTIPLE CHOICE

1. If two charges repel each other, the two charges must be
   a. positive and positive.
   b. positive and negative.
   c. negative and negative.
   d. Either (a) or (c)

   ANS: D   DIF: I   OBJ: 1-1.1

2. A device that can convert chemical energy to electrical energy is a
   a. lightning rod.
   b. cell.
   c. light bulb.
   d. All of the above

   ANS: B   DIF: I   OBJ: 1-2.1

3. Which of the following wires has the LOWEST resistance?
   a. a short, thick copper wire at 25°C
   b. a long, thick copper wire at 35°C
   c. a long, thin copper wire at 35°C
   d. a short, thick iron wire at 25°C

   ANS: A   DIF: I   OBJ: 1-3.2

4. An object becomes charged when the atoms in the object gain or lose
   a. protons.
   b. neutrons.
   c. electrons.
   d. All of the above

   ANS: C   DIF: I   OBJ: 1-1.2

5. A device used to protect buildings from electrical fires is a(n)
   a. electric meter.
   b. circuit breaker.
   c. fuse.
   d. Both (b) and (c)

   ANS: D   DIF: I   OBJ: 1-4.3

Holt Science and Technology
Copyright © by Holt, Rinehart and Winston. All rights reserved.

6. In order to produce a current from a cell, the electrodes of the cell must
   a. have a potential difference.
   b. be in a liquid.
   c. be exposed to light.
   d. be at two different temperatures.

   ANS: A   DIF: I   OBJ: 1-2.1

7. What type of current comes from the outlets in your home?
   a. direct current
   b. alternating current
   c. electric discharge
   d. static electricity

   ANS: B   DIF: I   OBJ: 1-3.4

8. Which of the following would LOWER the electrical resistance of a wire?
   a. making the wire thinner
   b. increasing the wire's length
   c. lowering the temperature of the wire
   d. using denser material for the wire

   ANS: C   DIF: I   OBJ: 1-3.2

9. If you rub a glass rod with a piece of silk, the rod becomes positively charged. This means that
   a. friction destroyed electrons in the rod.
   b. the silk has become negatively charged.
   c. protons have moved to the rod.
   d. glass attracts more protons.

   ANS: B   DIF: I   OBJ: 1-1.2

10. Which of the following is NOT an insulator?
    a. air
    b. water
    c. glass
    d. wood

    ANS: B   DIF: I   OBJ: 1-1.3

11. If you bring a charged object near an electrically neutral surface without allowing the object to touch the surface, the charges in the surface are rearranged by
    a. friction.
    b. induction.
    c. convection.
    d. conduction.

    ANS: B   DIF: I   OBJ: 1-1.2

12. In the United States, electrical circuits in homes and businesses
    a. use direct current.
    b. connect outlets and lights in series.
    c. normally receive 120 V.
    d. All of the above

    ANS: C   DIF: I   OBJ: 1-4.1

13. All matter is composed of very small particles called
    a. photons.
    b. molecules.
    c. atoms.
    d. elements.

    ANS: C   DIF: I   OBJ: 1-1.1

14. The law of electric charges states that
    a. every action has an equal and opposite reaction.
    b. charged objects that move produce electric current.
    c. the energy of charges and mass are interchangeable.
    d. like charges repel and opposite charges attract.

    ANS: D　　　DIF: I　　　OBJ: 1-1.1

15. Objects that have opposite charges
    a. are attracted to each other, and the force between the objects pulls them together.
    b. are attracted to each other, and the force between the objects pushes them apart.
    c. are repelled by each other, and the force between the objects pulls them together.
    d. are repelled by each other, and the force between the objects pushes them apart.

    ANS: A　　　DIF: I　　　OBJ: 1-1.1

16. Objects that have the same charge
    a. are attracted to each other, and the force between the objects pulls them together.
    b. are attracted to each other, and the force between the objects pushes them apart.
    c. are repelled by each other, and the force between the objects pulls them together.
    d. are repelled by each other, and the force between the objects pushes them apart..

    ANS: D　　　DIF: I　　　OBJ: 1-1.1

17. A region around a charged particle that can exert a force on another charged particle is called an
    a. electric field.　　　c. electric charge.
    b. electric force.　　　d. electric pulse.

    ANS: A　　　DIF: I　　　OBJ: 1-1.1

18. The strength of the electric force depends on the
    a. size of the objects.　　　c. the distance between the charges.
    b. size of the charges.　　　d. Both (b) and (c)

    ANS: D　　　DIF: II　　　OBJ: 1-1.1

19. If an electron is in the electric field of a proton, the electron is
    a. attracted to the proton by the electric force exerted on it.
    b. repelled by the proton by the electric force exerted on it.
    c. no longer able to move because the electric forces cancel.
    d. None of the above

    ANS: A　　　DIF: I　　　OBJ: 1-1.1

20. Objects can become charged by
    a. friction.　　　c. induction.
    b. conduction.　　　d. All of the above

    ANS: D　　　DIF: I　　　OBJ: 1-1.2

21. Charging by ____ occurs when electrons are transferred from one object to another by direct contact.
    a. reduction
    b. conduction
    c. induction
    d. friction

    ANS: B   DIF: I   OBJ: 1-1.2

22. When you rub a balloon on your hair, the balloon becomes charged by
    a. friction.
    b. conduction.
    c. induction.
    d. reduction.

    ANS: A   DIF: I   OBJ: 1-1.2

23. On a dry day, you can build up charge by shuffling your feet on a carpet. This is an example of charging by
    a. friction.
    b. conduction.
    c. induction.
    d. reduction.

    ANS: A   DIF: I   OBJ: 1-1.2

24. A negatively charged balloon near a wall will build up charge in the wall by
    a. friction.
    b. conduction.
    c. induction.
    d. reduction.

    ANS: C   DIF: I   OBJ: 1-1.2

25. Charge
    a. can be created.
    b. can be destroyed.
    c. is conserved.
    d. All of the above

    ANS: C   DIF: I   OBJ: 1-1.1

26. If an uncharged piece of metal touches a positively charged glass rod, charge is transferred by
    a. friction.
    b. conduction.
    c. reduction.
    d. induction.

    ANS: B   DIF: I   OBJ: 1-1.2

27. Plastic wrap clings to food containers because the wrap has become charged by
    a. friction.
    b. reduction.
    c. induction.
    d. conduction.

    ANS: A   DIF: I   OBJ: 1-1.2

28. A material through which charges can move easily is called a(n)
    a. conductor.
    b. insulator.
    c. inductor.
    d. reductor.

    ANS: A   DIF: I   OBJ: 1-1.3

Holt Science and Technology
Copyright © by Holt, Rinehart and Winston. All rights reserved.

29. A material in which charges cannot move easily is called a(n)
    a. conductor.
    b. insulator.
    c. inductor.
    d. reductor.

    ANS: B   DIF: I   OBJ: 1-1.3

30. The buildup of charges on an object is called
    a. an electric force.
    b. an electric field.
    c. static electricity.
    d. an electric charge.

    ANS: C   DIF: I   OBJ: 1-1.4

31. Clothes stick together when you pull them out of the dryer because
    a. clothing is a conductor.
    b. clothing is an inductor.
    c. they are not charged.
    d. of static electricity.

    ANS: D   DIF: I   OBJ: 1-1.4

32. The loss of static electricity as charges move off an object is called
    a. electric charge.
    b. electric discharge.
    c. electrical induction.
    d. electrical conduction.

    ANS: B   DIF: I   OBJ: 1-1.4

33. Lightning is an example of
    a. electric discharge.
    b. electric charge.
    c. electrical induction.
    d. electrical conduction.

    ANS: A   DIF: I   OBJ: 1-1.4

34. Objects that are in contact with the Earth are
    a. unable to conduct electric charge.
    b. insulators of electric charge.
    c. grounded.
    d. All of the above

    ANS: C   DIF: I   OBJ: 1-1.4

35. A(n) ____ produces an electric current by converting chemical energy into electrical energy.
    a. electroscope.
    b. fuse
    c. cell
    d. thermocouple

    ANS: C   DIF: I   OBJ: 1-2.1

36. A device that is made of several cells and produces an electric current by converting chemical energy into electrical energy is called
    a. an electroscope.
    b. a battery.
    c. a thermostat.
    d. a thermocouple.

    ANS: B   DIF: I   OBJ: 1-2.1

37. Every cell contains a mixture of chemicals that
    a. can conduct a current.
    b. can induct a current.
    c. acts as an insulator.
    d. is grounded.

    ANS: A   DIF: I   OBJ: 1-2.1

38. In a cell, electrolytes are
    a. inductors.
    b. conductors.
    c. insulators.
    d. grounded.

    ANS: B   DIF: I   OBJ: 1-2.1

39. The part of a cell through which charges enter or exit is the
    a. electrode.
    b. electrolyte.
    c. electrolytic converter.
    d. electrolytic inverter.

    ANS: A   DIF: I   OBJ: 1-2.1

40. Cells are made of
    a. a conducting wire and a grounded wire.
    b. an electrolyte and a pair of electrodes.
    c. a conductor and an insulator.
    d. None of the above

    ANS: B   DIF: I   OBJ: 1-2.1

41. When a zinc electrode and a copper electrode are dipped in a liquid electrolyte, the zinc electrode becomes negatively charged because a chemical reaction leaves extra
    a. protons on the zinc electrode.
    b. protons on the copper electrode.
    c. electrons on the zinc electrode.
    d. electrons on the copper electrode.

    ANS: C   DIF: I   OBJ: 1-2.1

42. When a zinc electrode and a copper electrode are dipped in a liquid electrolyte, the copper electrode becomes positively charged because a chemical reaction causes
    a. protons to be pulled off the zinc electrode.
    b. protons to be pulled off the copper electrode.
    c. electrons to be pulled off the zinc electrode.
    d. electrons to be pulled off the copper electrode.

    ANS: D   DIF: I   OBJ: 1-2.1

43. A zinc electrode and a copper electrode joined by a wire and dipped in an electrolyte is an example of a
    a. dry cell.
    b. wet cell.
    c. battery.
    d. photocell.

    ANS: B   DIF: I   OBJ: 1-2.1

44. The flow of electric charges is called
    a. electric current.
    b. static electricity.
    c. electric force.
    d. electric field.

    ANS: A DIF: I OBJ: 1-2.2

45. ____ is the energy per unit charge and is expressed in volts.
    a. Potential charge
    b. Potential current
    c. Potential force
    d. Potential difference

    ANS: D DIF: I OBJ: 1-2.2

46. A difference in charge between two electrodes in a cell causes a ____ between the electrodes.
    a. potential charge
    b. potential problem
    c. potential force
    d. potential difference

    ANS: D DIF: I OBJ: 1-2.2

47. The greater the potential difference,
    a. the lesser the current.
    b. the greater the current.
    c. the lesser the static electricity.
    d. the greater the static electricity.

    ANS: B DIF: I OBJ: 1-2.2

48. The part of a solar panel that converts light into electrical energy is called a
    a. cell.
    b. photocell.
    c. battery.
    d. thermocouple.

    ANS: B DIF: I OBJ: 1-2.3

49. Thermal energy can be converted into electrical energy by a
    a. thermostat.
    b. thermosphere.
    c. thermometer.
    d. thermocouple.

    ANS: D DIF: I OBJ: 1-2.3

50. How does a photocell convert light into electrical energy?
    a. Light contains electrons that are captured by a photocell.
    b. Light heats a photocell, and the difference in temperature creates an electric current.
    c. Light strikes silicon atoms in a photocell, ejecting electrons from the atoms.
    d. Light has mass, and the difference in mass on the photocell causes an electric current.

    ANS: C DIF: I OBJ: 1-2.3

51. How does a thermocouple convert thermal energy into electrical energy?
    a. Two wires of different metals are joined into a loop. The difference in the metals creates an electric current.
    b. Two wires of different metals are joined into a loop. The temperature difference within the loop creates an electric current.
    c. Two wires of different metals are joined into a loop. The chemical reaction within the loop creates an electric current.
    d. A single wire loop is heated, and the heat moves electrons in the wire.

    ANS: B   DIF: I   OBJ: 1-2.3

52. The rate at which charge passes a given point is called
    a. current.
    b. static electricity.
    c. potential difference.
    d. an ampere.

    ANS: A   DIF: I   OBJ: 1-3.1

53. Why does a light come on instantly when you flip on a light switch?
    a. Electrons move from one end of the wire to the other at the speed of light.
    b. Charges flow at the speed of light.
    c. An electric field sets up in the wire at nearly the speed of light.
    d. All of the above

    ANS: C   DIF: I   OBJ: 1-3.1

54. Charges flow in the same direction in
    a. alternating current.
    b. direct current.
    c. all types of current.
    d. static electricity.

    ANS: B   DIF: I   OBJ: 1-3.1

55. Charges continually switch from flowing in one direction to flowing in the reverse direction in
    a. alternating current.
    b. static electricity.
    c. all types of current.
    d. direct current.

    ANS: A   DIF: I   OBJ: 1-3.1

56. ____ is used in homes because it is more practical for transferring electrical energy.
    a. Photonic current
    b. Thermic current
    c. Alternating current
    d. Direct current

    ANS: C   DIF: I   OBJ: 1-3.1

57. Opposition to the flow of electric charge is called
    a. voltage.
    b. current.
    c. potential difference.
    d. resistance.

    ANS: D   DIF: I   OBJ: 1-3.2

Holt Science and Technology
Copyright © by Holt, Rinehart and Winston. All rights reserved.

58. Examine the wires below and answer the question that follows.

   Wire A

   Wire B

   Which wire has the greatest resistance?
   a. Wire A
   b. Wire B
   c. They both have the same resistance.
   d. It cannot be determined from the information given.

   ANS: B   DIF: II   OBJ: 1-3.2

59. Examine the wires below and answer the question that follows.

   Wire A

   Wire B

   Which wire has the greatest resistance?
   a. Wire A
   b. Wire B
   c. They both have the same resistance.
   d. It cannot be determined from the information given.

   ANS: A   DIF: II   OBJ: 1-3.2

60. Resistance is affected by
   a. temperature.   c. length.
   b. thickness.   d. All of the above

   ANS: D   DIF: I   OBJ: 1-3.2

61. In general, the resistance of metals
   a. increases as temperature increases.   c. decreases as temperature increases.
   b. increases as temperature decreases.   d. is the same regardless of temperature.

   ANS: A   DIF: I   OBJ: 1-3.2

Holt Science and Technology
Copyright © by Holt, Rinehart and Winston. All rights reserved.

62. Which of the following is the correct expression for Ohm's law?
    a. $V = \dfrac{I}{R}$
    b. $R = \dfrac{I}{V}$
    c. $I = \dfrac{V}{R}$
    d. None of the above

    ANS: C   DIF: I   OBJ: 1-3.3

63. Assume that your body has a resistance of 1,000,000Ω. What would the potential difference be across your body to produce a current of 0.001 A, which would cause a tingling feeling?
    a. 1 V
    b. 10 V
    c. 100 V
    d. 1,000 V

    ANS: D   DIF: II   OBJ: 1-3.3

64. Assume that your body has a resistance of 1,000,000Ω. What would the potential difference be across your body to produce a current of 0.015 A, a fatal amount of current?
    a. 1.5 V
    b. 15 V
    c. 150 V
    d. 1,500 V

    ANS: D   DIF: II   OBJ: 1-3.3

65. The rate at which electrical energy is used to do work is called
    a. electric current.
    b. electrical potential.
    c. electric power.
    d. static electricity.

    ANS: C   DIF: I   OBJ: 1-3.4

66. Which of the following is the correct expression relating power, voltage, and current?
    a. $I = \dfrac{V}{P}$
    b. $P = \dfrac{V}{I}$
    c. $P = \dfrac{I}{V}$
    d. $P = V \times I$

    ANS: D   DIF: I   OBJ: 1-3.4

67. Which of the following bulbs would burn the brightest?
    a. 40 W
    b. 65 W
    c. 100 W
    d. 120 W

    ANS: D   DIF: I   OBJ: 1-3.4

68. Which of the following is the correct expression relating electrical energy, power, and time?
    a. $E = \dfrac{t}{P}$
    b. $E = P \times t$
    c. $P = \dfrac{t}{E}$
    d. $P = E \times t$

    ANS: B   DIF: I   OBJ: 1-3.4

69. How much electrical energy is used by a 200-watt color television that stays on for 4 hours?
    a. 0.2 kWh
    b. 0.4 kWh
    c. 0.6 kWh
    d. 0.8 kWh

    ANS: D   DIF: I   OBJ: 1-3.4

70. A device that uses electrical energy to do work is called a
    a. circuit.
    b. load.
    c. series circuit.
    d. parallel circuit.

    ANS: B   DIF: I   OBJ: 1-4.1

71. All circuits include
    a. an energy source, a load, and wires.
    b. an energy source, a resistor, and a battery.
    c. a battery, a light bulb, and a switch.
    d. a battery, wires, and a switch.

    ANS: A   DIF: I   OBJ: 1-4.1

72. When a switch is closed, two pieces of conducting material
    a. touch, allowing the electric charges to flow through the circuit.
    b. touch, preventing the electric charges from flowing through the circuit.
    c. do not touch, allowing the electric charges to flow through the circuit.
    d. do not touch, preventing the electric charges from flowing through the circuit.

    ANS: A   DIF: I   OBJ: 1-4.1

73. When a switch is open, two pieces of conducting material
    a. touch, allowing the electric charges to flow through the circuit.
    b. do not touch, allowing the electric charges to flow through the circuit.
    c. touch, preventing the electric charges from flowing through the circuit.
    d. do not touch, preventing the electric charges from flowing through the circuit.

    ANS: D   DIF: I   OBJ: 1-4.1

74. A circuit in which all parts are connected in a single loop is called a(n)
    a. open circuit.
    b. open load.
    c. series circuit.
    d. parallel circuit.

    ANS: C   DIF: I   OBJ: 1-4.2

75. A circuit in which different loads are located on separate branches is called a(n)
    a. open circuit.
    b. open load.
    c. series circuit.
    d. parallel circuit.

    ANS: D   DIF: I   OBJ: 1-4.2

76. All the loads in a series circuit
    a. use the same voltage.
    b. share the same current.
    c. have the same resistance.
    d. have the same power.

    ANS: B   DIF: I   OBJ: 1-4.2

77. Suppose you have four bulbs in a series circuit. If you were to add five more bulbs in series with these four, what would happen to the brightness of the bulbs?
    a. The bulbs would no longer glow.
    b. The bulbs would grow dimmer.
    c. The bulbs would grow brighter.
    d. The brightness would not change.

    ANS: B   DIF: I   OBJ: 1-4.2

78. Suppose you have four bulbs in a parallel circuit. If you were to add five more bulbs in parallel with these four, what would happen to the brightness of the bulbs?
    a. The bulbs would no longer glow.
    b. The bulbs would grow brighter.
    c. The bulbs would grow dimmer.
    d. The brightness would not change.

    ANS: D   DIF: I   OBJ: 1-4.2

79. All the loads in a parallel circuit
    a. use the same voltage.
    b. share the same current.
    c. have the same resistance.
    d. have the same power.

    ANS: A   DIF: I   OBJ: 1-4.2

80. What is meant by a short circuit?
    a. Charges bypass the loads in the circuit.
    b. The power source no longer works.
    c. The electric field is no longer in the wire.
    d. The circuit is not long enough to power the load.

    ANS: A   DIF: I   OBJ: 1-4.3

81. Which of the following safety features are found in circuits in homes?
    a. insulation around wires
    b. fuses
    c. circuit breakers
    d. All of the above

    ANS: D   DIF: I   OBJ: 1-4.3

82. Which of the following might cause a circuit to fail?
    a. connecting a series circuit to a parallel circuit
    b. plugging too many devices into one outlet
    c. failing to press the TEST button on a circuit breaker box
    d. All of the above

    ANS: B   DIF: I   OBJ: 1-4.3

83. What can happen if all the loads are bypassed in a short circuit?
    a. The resistance of the circuit drops.
    b. The current increases.
    c. A fire could start.
    d. All of the above

    ANS: D   DIF: I   OBJ: 1-4.3

COMPLETION

1. A _____ converts chemical energy into electrical energy. (battery or photocell)

    ANS: battery   DIF: I   OBJ: 1-2.1

2. Charges flow easily in a(n) _____. (insulator or conductor)

    ANS: conductor   DIF: I   OBJ: 1-1.3

3. _____ is the opposition to the flow of electric charge. (Resistance or Electric power)

    ANS: Resistance   DIF: I   OBJ: 1-3.2

4. A _____ is a complete, closed path through which charges flow. (load or circuit)

    ANS: circuit   DIF: I   OBJ: 1-4.1

5. Lightning is a form of _____. (static electricity or electric discharge)

    ANS: electric discharge   DIF: I   OBJ: 1-1.4

6. _____ is the rate at which charge passes a given point. (Current or Induction)

    ANS: Current   DIF: I   OBJ: 1-3.1

7. The buildup of electric charges on an object is called _____. (electric discharge or static electricity)

    ANS: static electricity   DIF: I   OBJ: 1-1.4

8. In a _____, the current passing through each load is the same. (series circuit or parallel circuit)

    ANS: series circuit   DIF: I   OBJ: 1-4.2

9. _____ are used to monitor the temperature of car engines, furnaces, and ovens. (Thermocouples or Photocells)

    ANS: Thermocouples   DIF: I   OBJ: 1-2.3

Holt Science and Technology
Copyright © by Holt, Rinehart and Winston. All rights reserved.

10. A _____ of 12 V exists between the poles of a common car battery. (resistance or potential difference)

   ANS: potential difference   DIF: I   OBJ: 1-2.2

11. The force between charged objects is an _____.

   ANS: electric force   DIF: I   OBJ: 1-1.1

12. Charging by _____ occurs when charges in an uncharged object are rearranged without directly contacting a charged object.

   ANS: induction   DIF: I   OBJ: 1-1.2

SHORT ANSWER

1. Describe how an object is charged by friction.

   ANS:
   When objects are rubbed together, friction transfers electrons between them. Objects losing electrons become positively charged, while objects gaining electrons become negatively charged.

   DIF: I   OBJ: 1-1.2

2. Compare charging by conduction and induction.

   ANS:
   Charging by *conduction* involves direct contact between objects, while charging by *induction* does not. With *conduction*, electrons flow between objects that are touching each other. With *induction*, electrons migrate within objects so that the sides of the objects nearest each other are oppositely charged.

   DIF: I   OBJ: 1-1.2

3. Suppose you are conducting experiments using an electroscope. You touch an object to the top of the electroscope, the metal leaves spread apart, and you determine that the object has a charge. However, you cannot determine the type of charge (positive or negative) the object has. Explain why not.

   ANS:
   Like charges repel, but repulsion does not indicate what charge—positive or negative—is involved. Therefore, the electroscope will behave the same, regardless of an object's charge.

   DIF: II   OBJ: 1-1.1

Holt Science and Technology
Copyright © by Holt, Rinehart and Winston. All rights reserved.

4. What is static electricity? Give an example of static electricity.

   ANS:
   Static electricity is the buildup of electric charge on an object. Examples include clothes sticking together after being machine dried and hair sticking up after being brushed.

   DIF: I   OBJ: 1-1.4

5. How is the shock you receive from a metal doorknob similar to a bolt of lightning?

   ANS:
   Both the shock you receive from a doorknob and a bolt of lightning are examples of electric discharge.

   DIF: I   OBJ: 1-1.4

6. When you use an electroscope, you touch a charged object to a metal rod that is held in place by a rubber stopper. Why is it important to touch the object to the metal rod and NOT to the rubber stopper?

   ANS:
   The metal rod is a conductor, so electrons can easily move down the rod to the metal leaves or up the rod from the metal leaves and confirm any charge present. The rubber stopper is an insulator. Electrons will not move through rubber, so the electroscope will not perform as intended.

   DIF: II   OBJ: 1-1.3

7. Name the parts of a cell, and explain how they work together to produce an electric current.

   ANS:
   A cell is made of an electrolyte and two electrodes. Chemical reactions in the electrolyte leave extra electrons on one electrode and strip them from the other. If the charged electrodes are connected with a wire, electric charges will flow between them.

   DIF: I   OBJ: 1-2.1

8. How do the currents produced by a 1.5 V flashlight cell and a 12 V car battery compare?

   ANS:
   Under equal conditions, the current from a 12 V car battery is greater than the current from a 1.5 V flashlight cell.

   DIF: I   OBJ: 1-2.2

9. Why do you think some solar calculators contain batteries?

   ANS:
   Some solar calculators contain batteries as a backup power source, for when there is insufficient light.

   DIF: II   OBJ: 1-2.3

10. What is electric current?

    ANS:
    Electric current is a continuous flow of charge caused by the motion of electrons.

    DIF: I   OBJ: 1-3.1

11. How does increasing the voltage affect the current?

    ANS:
    Current depends partially on voltage, so increasing the voltage increases the current.

    DIF: I   OBJ: 1-3.3

12. How does an electric power company calculate electrical energy from electric power?

    ANS:
    Electric power companies calculate electrical energy by multiplying the power in kilowatts by the time in hours.

    DIF: I   OBJ: 1-3.4

13. Which wire would have the lowest resistance: a long, thin wire at a high temperature or a short, thick copper wire at a low temperature?

    ANS:
    A short, thick copper wire at a low temperature would have a lower resistance than a long, thin iron wire at a high temperature.

    DIF: II   OBJ: 1-3.2

14. Use Ohm's law to find the voltage needed to produce a current of 3 A in a device with a resistance of 9 Ω.

    ANS:
    $V = I \times R = 3 \text{ A} \times 9 \text{ } \Omega = 27 \text{ V}$

    DIF: II   OBJ: 1-3.3

Holt Science and Technology
Copyright © by Holt, Rinehart and Winston. All rights reserved.

15. Name and describe the three essential parts of a circuit.

    ANS:
    The *energy source* provides electrical energy. The *load* is any device that uses the electrical energy to do work. *Wires* connect the energy source to the load.

    DIF: I   OBJ: 1-4.1

16. Why are switches useful in a circuit?

    ANS:
    Switches are useful because they allow you to turn electrical devices on and off without removing them from the circuit.

    DIF: I   OBJ: 1-4.1

17. What is the difference between series and parallel circuits?

    ANS:
    All parts of a *series* circuit are connected in a single loop. In a *parallel* circuit, the loads are attached to the circuit on different branches.

    DIF: I   OBJ: 1-4.2

18. How do fuses and circuit breakers protect your home against electrical fires?

    ANS:
    Fuses and circuit breakers create gaps in a circuit when the current gets too high, preventing charges from flowing. "Breaking" the circuit prevents overheating and fires.

    DIF: I   OBJ: 1-4.3

19. Whenever you turn on the portable heater in your room, the circuit breaker for the circuit in your room opens and all the lights go out. Propose two possible reasons for why this occurs.

    ANS:
    One possible reason is that there is a short circuit in the portable heater. Another possible reason is that the heater, along with any other devices attached to the circuit, overloads the circuit and trips the breaker.

    DIF: III   OBJ: 1-4.3

20. Briefly explain the relationship between charge and force.

    ANS:
    Charge is a physical property. Objects with a positive or negative charge exert a *force* on other charged objects.

    DIF: I   OBJ: 1-1.1

Holt Science and Technology
Copyright © by Holt, Rinehart and Winston. All rights reserved.

21. Discuss the difference between a conductor and an insulator. Give an example of each.

    ANS:
    Charge moves easily in a *conductor* but has difficulty moving in an *insulator*. Most metals are conductors. Plastic, rubber, and glass are insulators.

    DIF: I   OBJ: 1-1.3

22. Explain what potential difference is and explain the difference between volts and potential difference.

    ANS:
    Electric charges flow between the electrodes of a cell or battery. *Potential difference* is the energy per unit charge. *Volts* are the units used to express the potential difference.

    DIF: I   OBJ: 1-2.2

23. Explain the difference between wet cells and dry cells.

    ANS:
    Wet cells contain a liquid electrolyte such as sulfuric acid. *Dry cells*, such as flashlight cells, contain solid or pastelike electrolytes.

    DIF: I   OBJ: 1-2.1

24. According to Ohm's law, what happens to the current if the voltage increases and the resistance stays constant?

    ANS:
    The current increases if the voltage increases and the resistance stays constant.

    DIF: I   OBJ: 1-3.3

25. If the current in a wire is 4 A, what is the ratio of the voltage applied to the wire to the wire's resistance in ohms?

    ANS:
    The ratio is 4:1.

    DIF: I   OBJ: 1-3.3

26. How do electric power companies keep track of how much electrical energy a household or business uses?

    ANS:
    Usually they use electric meters that record the kilowatt-hours of energy used by a household or a business.

    DIF: I   OBJ: 1-3.4

27. List and describe the three essential parts of a circuit.

    ANS:
    The three essential parts of a circuit are an energy source, a load, and wires. The *energy source* provides electrical energy to do work. The *load* is a device that uses the electrical energy to do work. Loads cause electrical energy to be converted to other forms of energy. The *wires* are used to connect all the other parts of the circuit together. Wires are made of conductors.

    DIF: I   OBJ: 1-4.1

28. Name the two factors that affect the strength of electric force, and explain how they affect electric force.

    ANS:
    One factor is the amount of the electric charge. The greater the charge is, the greater the force. The other factor is the distance between the charges. The closer the charges are to each other, the greater the force is.

    DIF: I   OBJ: 1-1.1

29. Describe how direct current differs from alternating current.

    ANS:
    The charges in direct current flow in one direction. In alternating current, the charges continually switch from flowing in one direction to flowing in the reverse direction.

    DIF: I   OBJ: 1-3.1

30. Use the following terms to create a concept map: *electric current, battery, charges, photocell, thermocouple, circuit, parallel circuit, series circuit*.

    ANS:
    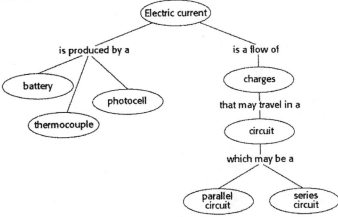

    DIF: I   OBJ: 1-4.2

31. Your science classroom was rewired over the weekend. On Monday, you notice that the electrician may have made a mistake. In order for the fish-tank bubbler to work, the lights in the room must be on. And if you want to use the computer, you must turn on the overhead projector. Describe what mistake the electrician made with the circuits in your classroom.

    ANS:
    The electrician must have wired the fish-tank bubbler in series with the lights and the computer in series with the overhead projector.

    DIF: II   OBJ: 1-4.2

32. You can make a cell using an apple, a strip of copper, and a strip of silver. Explain how you would construct the cell, and identify the parts of the cell. What type of cell is formed? Explain your answer.

    ANS:
    You would push the strip of copper and the strip of silver into the apple. The apple is the electrolyte, and the metal strips are the electrodes. Students may identify the cell as a dry cell because the apple is a solid or as a wet cell because the apple juice conducts the electric current.

    DIF: II   OBJ: 1-2.1

33. Your friend shows you a magic trick. She rubs a plastic comb with a piece of silk and holds the comb close to a stream of water. When the comb is close to the water, the water bends toward the comb. Explain how this trick works. (Hint: Think about how objects become charged.)

    ANS:
    When the comb is rubbed with a piece of silk, the comb is charged by friction. When the charged comb is held close to the stream of water, the charged comb induces a charge on the stream of water. The part of the stream closest to the comb has a charge opposite to that of the comb. Therefore, the stream is attracted by an electric force to the comb.

    DIF: II   OBJ: 1-1.2

34. What voltage is needed to produce a 6 A current through a resistance of 3 Ω?

    ANS:
    $V = I \times R = 6 \text{ A} \times 3 \text{ Ω} = 18 \text{ V}$

    DIF: II   OBJ: 1-3.3

35. Find the current produced when a voltage of 60 V is applied to a resistance of 15 Ω.

    ANS:
    $I = V \div R = 60 \text{ V} \div 15 \text{ Ω} = 4 \text{ A}$

    DIF: II   OBJ: 1-3.3

36. What is the resistance of an object if a voltage of 40 V produces a current of 5 A?

    ANS:
    $R = V \div I = 40 \text{ V} \div 5 \text{ A} = 8 \text{ } \Omega$

    DIF: II   OBJ: 1-3.3

37. Classify the objects in the illustration below as conductors or insulators.

    ANS:
    Conductors: tap water in glass, wrench, metal part of scissors, liquid soap in plastic bottle; *insulators*: basketball, glass, plastic bottle, plastic scissors handles, wooden table.

    DIF: II   OBJ: 1-1.3

38. What does Ohm's law tell us?

    ANS:
    Ohm's law gives the relationship between current, voltage, and resistance. It is summarized in the formula $I = \dfrac{V}{R}$; the current (I) in a circuit equals voltage (V) divided by resistance (R).

    DIF: I   OBJ: 1-3.3

39. How does lightning form?

    ANS:
    During thunderstorms, negative charges build up at the bottom of clouds. These negative charges induce a positive charge on the surface of the ground below. Lightning is a rapid electric discharge between a cloud and the oppositely charged ground or between oppositely charged parts of clouds.

    DIF: I   OBJ: 1-1.4

40. Why would it be dangerous to replace a blown fuse with a copper penny?

    ANS:
    The metal strip in the fuse melted because too much current was flowing through the circuit. If the fuse had not blown, the wires in the circuit would have overheated and might have caused a fire. Replacing the fuse with a penny would restore the excessive current, but the penny would not burn out; therefore, the wires would overheat and a fire might result.

    DIF: II   OBJ: 1-4.3

41. As an added safety precaution, would it be useful to insert two fuses in a circuit instead of one? Explain your answer.

    ANS:
    No; inserting two fuses would be useless because the circuit will already be broken if one fuse blows.

    DIF: II   OBJ: 1-4.3

42. Find the resistance of a circuit that draws 1.5 A when 3.0 V are applied. Show your work.

    ANS:
    By Ohm's law, $R = \dfrac{V}{I}$; thus, the resistance of the circuit is $\dfrac{3.0 \text{ V}}{1.5 \text{ A}} = 2 \, \Omega$.

    DIF: II   OBJ: 1-3.3

43. Use the diagram below to answer the following question.

    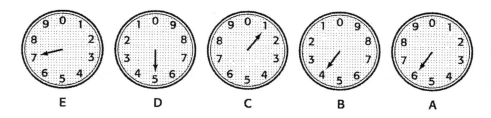

    E   D   C   B   A

    The power company measures electrical energy in kilowatt-hours (kWh). The meter in the picture above measures the number of kWh used in a building. The dials are arranged by place value: dial **A** has a value in ones, dial **B** in tens, dial **C** in hundreds, dial **D** in thousands, and dial **E** in ten-thousands. What is the combined reading, in kWh, of the meters above?

    ANS:
    The meter registers 75,146 kWh.

    DIF: II   OBJ: 1-3.4

44. Use the circuit diagram below to answer the following question.

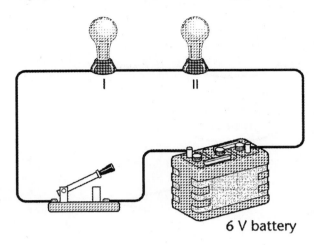

If one bulb in this circuit burns out, how will the other bulb be affected?

ANS:
If one bulb burns out, there will be an open circuit; therefore, the other bulb will also go out.

DIF: II OBJ: 1-4.2

45. Use the following terms to complete the concept map below: *electrolyte, car battery, flashlight battery, energy transfer, DC, electric current, AC.*

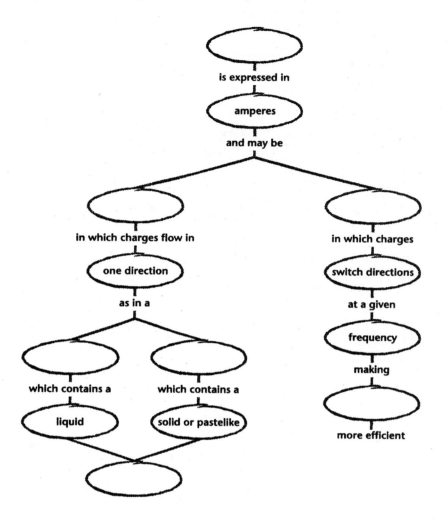

ANS:

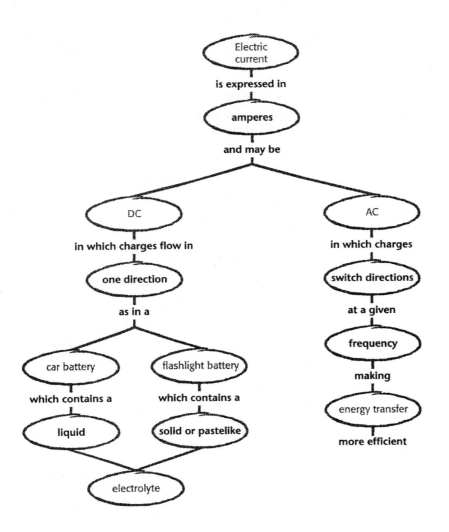

DIF: II OBJ: 1-2.1

46. When traveling to another country, you should always find out the voltage that is used in that country before you plug in an appliance. To understand the reason for this precaution, calculate the current that a laptop computer would draw from a 120 V outlet in the United States if the computer has a resistance of 40.0 Ω. Then, calculate the current that the same computer would draw if you plugged it into a 240 V outlet in the United Kingdom.

ANS:
$I = \dfrac{V}{R} = \dfrac{120 \text{ V}}{40 \text{ Ω}} = 3.0$ A in the United States

$I = \dfrac{V}{R} = \dfrac{240 \text{ V}}{40 \text{ Ω}} = 6.0$ A in the United Kingdom

DIF: II OBJ: 1-3.3

47. Why are fuses useful in household circuits?

    ANS:
    A fuse contains a thin strip of metal through which the charges for a circuit flow. If the current in the circuit is too high, the metal in the fuse warms up and melts, creating a gap in the circuit. This is useful because an overloaded circuit could cause a fire.

    DIF: II   OBJ: 1-4.3

Short Course N Chapter 2—Electromagnetism

MULTIPLE CHOICE

1. The region around a magnet in which magnetic forces can act is called the
   a. magnetic field.
   b. domain.
   c. pole.
   d. solenoid.

   ANS: A   DIF: I   OBJ: 2-1.1

2. An electric fan has an electric motor inside to change
   a. kinetic energy into electrical energy.
   b. thermal energy into electrical energy.
   c. electrical energy into thermal energy.
   d. electrical energy into kinetic energy.

   ANS: D   DIF: I   OBJ: 2-2.3

3. The marked end of a compass needle always points directly to
   a. Earth's geographic South Pole.
   b. Earth's geographic North Pole.
   c. a magnet's south pole.
   d. a magnet's north pole.

   ANS: C   DIF: I   OBJ: 2-1.4

4. A device that increases the voltage of an alternating current is called a(n)
   a. electric motor.
   b. galvanometer.
   c. step-up transformer.
   d. step-down transformer.

   ANS: C   DIF: I   OBJ: 2-3.3

5. The magnetic field of a solenoid can be increased by
   a. adding more loops.
   b. increasing the current.
   c. putting an iron core inside the coil to make an electromagnet.
   d. All of the above

   ANS: D   DIF: I   OBJ: 2-2.2

6. What do you end up with if you cut a magnet in half?
   a. one north-pole piece and one south-pole piece
   b. two unmagnetized pieces
   c. two pieces, each with a north pole and a south pole
   d. two north-pole pieces

   ANS: C   DIF: I   OBJ: 2-1.2

7. The magnetic effects of a bar magnet are strongest near the ____.
   a. center
   b. top
   c. ends
   d. bottom

   ANS: C   DIF: I   OBJ: 2-1.1

8. Earth's magnetic field
   a. is produced by a giant magnet at the center of the Earth.
   b. is produced by the movement of electric charges in Earth's core.
   c. does not change over time.
   d. bends outward at the magnetic poles.

   ANS: B   DIF: I   OBJ: 2-1.4

9. ____ discovered the relationship between electricity and magnetism.
   a. William Gilbert
   b. Michael Faraday
   c. Joseph Henry
   d. Hans Christian Oersted

   ANS: D   DIF: I   OBJ: 2-2.1

10. A galvanometer does not include a(n)
    a. electromagnet.
    b. battery.
    c. pointer.
    d. permanent magnet.

    ANS: B   DIF: I   OBJ: 2-2.3

11. Which of the following is NOT true about an electromagnet?
    a. It can be strong enough to levitate a train.
    b. The current may be turned on or off.
    c. Its strength is reduced by adding more loops.
    d. It consists of an iron core and a current-carrying wire.

    ANS: C   DIF: I   OBJ: 2-2.2

12. Any material that attracts iron or materials containing iron is called a
    a. solenoid.
    b. magnet.
    c. superconductor.
    d. semiconductor.

    ANS: B   DIF: I   OBJ: 2-1.1

13. All magnets
    a. have two poles.
    b. exert forces.
    c. are surrounded by a magnetic field.
    d. All of the above

    ANS: D   DIF: I   OBJ: 2-1.1

14. The magnetic effects of a bar magnet are
    a. evenly distributed throughout a magnet.
    b. distributed randomly in a magnet.
    c. concentrated near the ends.
    d. None of the above

    ANS: C   DIF: I   OBJ: 2-1.1

15. Sometimes the magnetic strip on a credit card becomes demagnetized. This strip is most likely a(n)
    a. permanent magnet.
    b. temporary magnet.
    c. electromagnet.
    d. geomagnet.

    ANS: B   DIF: II   OBJ: 2-1.3

16. If you attach a magnet to a string so that the magnet is free to rotate, you will see that one end of the magnet will point
    a. north.
    b. southwest.
    c. east.
    d. west.

    ANS: A   DIF: I   OBJ: 2-1.1

17. The pole of a magnet that points to the north is called the magnet's
    a. north pole.
    b. south pole.
    c. east pole.
    d. west pole.

    ANS: A   DIF: I   OBJ: 2-1.1

18. The pole of a magnet that points to the south is called the magnet's
    a. north pole.
    b. south pole.
    c. east pole.
    d. west pole.

    ANS: B   DIF: I   OBJ: 2-1.1

19. The Earth has neither an east pole nor a west pole because
    a. the Earth is a sphere and has no poles.
    b. the Earth is a sphere and has only one pole called an axis.
    c. the Earth is like a bar magnet, which only has two poles.
    d. None of the above

    ANS: C   DIF: I   OBJ: 2-1.4

20. A compass needle is
    a. a fixed magnet.
    b. a fixed nonmagnetic piece of metal.
    c. a magnet that is free to rotate.
    d. a nonmagnetic piece of metal that is free to rotate.

    ANS: C   DIF: I   OBJ: 2-1.1

21. Magnetic poles always occur
    a. alone.
    b. in pairs.
    c. in threes.
    d. in fours.

    ANS: B   DIF: I   OBJ: 2-1.1

22. The force of repulsion or attraction between the poles of magnets is called the
    a. Coulomb force.
    b. Foucault force.
    c. magnetic force.
    d. gravitational force.

    ANS: C   DIF: I   OBJ: 2-1.1

23. The magnetic force will push magnets apart if you hold the
    a. north poles of two magnets close together.
    b. south poles of two magnets close together.
    c. north pole of one magnet near the south pole of another magnet.
    d. Both (a) and (b)

    ANS: D   DIF: I   OBJ: 2-1.1

24. The magnetic force will pull magnets together if you hold the
    a. north pole of one magnet near the south pole of another magnet.
    b. south poles of two magnets close together.
    c. north poles of two magnets close together.
    d. Both (a) and (b)

    ANS: A   DIF: I   OBJ: 2-1.1

25. Magnetic poles are similar to electric charges in that
    a. like poles repel and opposite poles attract.
    b. the magnetic force is equal to the electric force.
    c. the number of magnetic domains responsible for the poles is conserved.
    d. the mass of the magnetized particle is conserved.

    ANS: A   DIF: I   OBJ: 2-1.1

26. Which of the following describes the magnetic field of a bar magnet?
    a. the region in which magnetic forces can act
    b. strongest at the poles of the magnet
    c. its shape can be shown by magnetic field lines
    d. All of the above

    ANS: D   DIF: I   OBJ: 2-1.1

27. The magnetic field around a bar magnet can be modeled by drawing
    a. curved lines from the north pole of the magnet to the south pole.
    b. curved lines near the south pole of the magnet.
    c. straight lines from the north pole and from the south pole.
    d. circular lines around the entire magnet.

    ANS: A   DIF: I   OBJ: 2-1.1

Holt Science and Technology
Copyright © by Holt, Rinehart and Winston. All rights reserved.

28. If you were to sprinkle iron filings over a bar magnet, where would they fall?
    a. between the field lines
    b. along the field lines
    c. only on top of the bar magnet
    d. only at the ends of the bar magnet

    ANS: B   DIF: II   OBJ: 2-1.1

29. Whether a material is magnetic depends on the ____ in the material.
    a. molecules   c. number of neutrons
    b. atoms   d. number of protons

    ANS: B   DIF: I   OBJ: 2-1.2

30. ____ produce the magnetic fields that can give an atom a north and a south pole.
    a. Moving protons   c. Moving electrons
    b. Moving neutrons   d. Moving magnetospheres

    ANS: C   DIF: I   OBJ: 2-1.2

31. Why are most materials not magnetic?
    a. The electrons stop moving.
    b. Only iron atoms can have magnetic fields.
    c. The magnetic fields of the individual atoms cancel each other out.
    d. Most atoms don't have magnetic fields.

    ANS: C   DIF: I   OBJ: 2-1.2

32. What makes materials magnetic?
    a. The atoms in these materials are magnetized by moving electrons.
    b. The atoms in these materials are randomly arranged.
    c. The atoms in a domain are arranged so that the north and south poles of all the atoms line up.
    d. All of the above

    ANS: C   DIF: I   OBJ: 2-1.2

Holt Science and Technology
Copyright © by Holt, Rinehart and Winston. All rights reserved.

Examine the magnetic field lines around the magnets below, and answer the questions that follow:

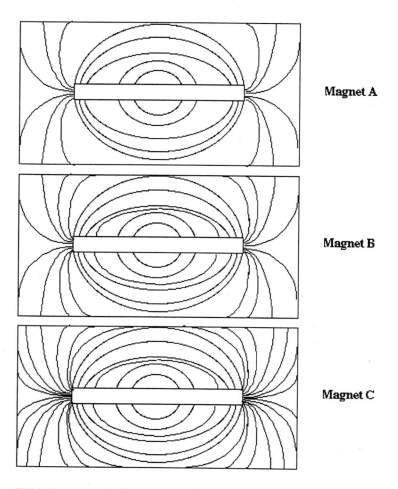

Magnet A

Magnet B

Magnet C

33. Which magnet is the strongest?
    a. **Magnet A**
    b. **Magnet B**
    c. **Magnet C**
    d. All magnets have the same strength.

    ANS: C   DIF: II   OBJ: 2-1.1

34. Which magnet is the weakest?
    a. **Magnet A**
    b. **Magnet B**
    c. **Magnet C**
    d. All magnets have the same strength.

    ANS: A   DIF: II   OBJ: 2-1.1

35. If the north pole of **Magnet A** is to the left, the arrows of the field lines should point
    a. from right to left.
    b. from left to right.
    c. clockwise (from right to left on the top and left to right on the bottom).
    d. counterclockwise (from left to right on the top and right to left on the bottom).

    ANS: B   DIF: II   OBJ: 2-1.1

Holt Science and Technology
Copyright © by Holt, Rinehart and Winston. All rights reserved.

The diagrams below illustrate magnetic domains within two samples of a metal. Unshaded areas correspond to one type of pole, and darkly shaded areas correspond to the opposite type of pole.

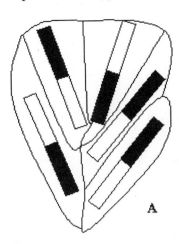
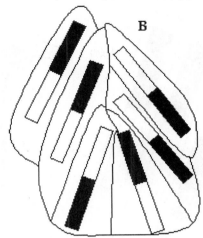

36. Which sample is magnetized?
    a. **A**
    b. **B**
    c. Both **A** and **B**
    d. Neither **A** nor **B**

    ANS: A   DIF: II   OBJ: 2-1.2

The diagrams below illustrate magnetic domains within two metallic substances. Unshaded areas correspond to one type of pole, and darkly shaded areas correspond to the opposite type of pole.

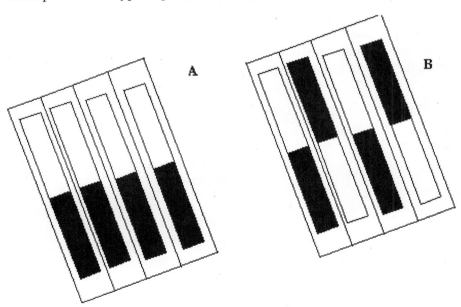

37. Which substance is magnetized?
    a. **A**
    b. **B**
    c. Both **A** and **B**
    d. Neither **A** nor **B**

    ANS: A   DIF: II   OBJ: 2-1.2

Holt Science and Technology
Copyright © by Holt, Rinehart and Winston. All rights reserved.

The diagrams below illustrate magnetic domains within two samples of a metal. Unshaded areas correspond to one type of pole, and darkly shaded areas correspond to the opposite type of pole.

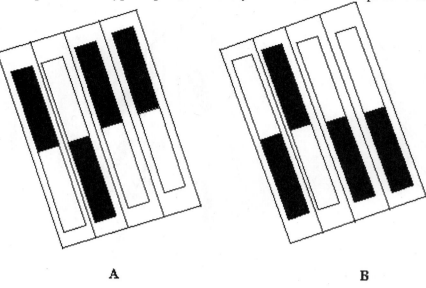

A   B

38. Which sample has the greatest magnetic field?
    a. A
    b. B
    c. Both A and B are equal in strength.
    d. Neither A nor B is magnetic.

    ANS: C   DIF: II   OBJ: 2-1.2

39. Once the domains of a magnet are
    a. aligned, they can become randomly oriented.
    b. aligned, they do not become randomly oriented.
    c. randomly oriented, they do not become aligned.
    d. None of the above

    ANS: A   DIF: I   OBJ: 2-1.2

40. A magnet can demagnetize if
    a. it is dropped.
    b. it is heated.
    c. it is struck very hard.
    d. All of the above

    ANS: D   DIF: I   OBJ: 2-1.2

41. A magnet made with iron, nickel, cobalt, or alloys of these metals is called a(n)
    a. electromagnet.
    b. ferromagnet.
    c. temporary magnet.
    d. magnetic domain.

    ANS: B   DIF: I   OBJ: 2-1.3

42. Permanent magnets
    a. cannot lose their magnetization.
    b. are difficult to magnetize.
    c. have randomly oriented domains.
    d. are easy to magnetize.

    ANS: B   DIF: I   OBJ: 2-1.3

Holt Science and Technology
Copyright © by Holt, Rinehart and Winston. All rights reserved.

43. Temporary magnets
    a. have randomly oriented domains.
    b. are difficult to magnetize.
    c. cannot lose their magnetization.
    d. are easy to magnetize.

    ANS: D   DIF: I   OBJ: 2-1.3

44. A magnet, usually with an iron core, produced by an electric current is a(n)
    a. electromagnet.
    b. ferromagnet.
    c. temporary magnet.
    d. permanent magnet.

    ANS: A   DIF: I   OBJ: 2-1.3

45. One of the most spectacular effects caused by the Earth's magnetic field is a curtain of light called
    a. an aurora.
    b. lightning.
    c. a sprite.
    d. an elf.

    ANS: A   DIF: I   OBJ: 2-1.4

46. Aurora australis is seen near the
    a. geographic North Pole.
    b. geographic South Pole.
    c. south magnetic pole.
    d. Both (b) and (c)

    ANS: D   DIF: I   OBJ: 2-1.4

47. Aurora borealis is seen near the
    a. geographic North Pole.
    b. geographic South Pole.
    c. north magnetic pole.
    d. Both (a) and (c)

    ANS: D   DIF: I   OBJ: 2-1.4

48. Earth's geographic North Pole is actually
    a. on the axis on which Earth rotates.
    b. near the magnetic south pole.
    c. near the magnetic north pole.
    d. Both (a) and (b)

    ANS: D   DIF: I   OBJ: 2-1.4

49. The direction of a magnetic field around a wire depends on
    a. the strength of the current.
    b. the direction of the current.
    c. the rate at which the current increases.
    d. All of the above

    ANS: B   DIF: I   OBJ: 2-2.1

50. If no electric current exists in the wire of a simple circuit, the compass needles around the wire will
    a. point in the same direction.
    b. point in a direction opposite the current.
    c. deflect in a counterclockwise direction.
    d. deflect in a clockwise direction.

    ANS: A   DIF: I   OBJ: 2-2.1

Holt Science and Technology
Copyright © by Holt, Rinehart and Winston. All rights reserved.

51. If the electric current in the wire of a simple circuit causes compass needles around the wire to deflect in a counterclockwise direction, when the current is reversed, the compass needles
    a. point in the same direction as the current.
    b. point in the direction opposite the current.
    c. continue pointing in a counterclockwise direction.
    d. deflect in a clockwise direction.

    ANS: D　　　DIF: II　　　OBJ: 2-2.1

52. If each of the following solenoids carries the same electric current, which one would produce the strongest magnetic field?
    a. a solenoid with a single loop
    b. a solenoid with 10 loops
    c. a solenoid with 100 loops
    d. a solenoid with 1,000 loops

    ANS: D　　　DIF: I　　　OBJ: 2-2.2

53. Four solenoids have the same number of loops. Which solenoid would produce the strongest magnetic field?
    a. the solenoid with 1 A of current
    b. the solenoid with 100 A of current
    c. the solenoid with 10 A of current
    d. the solenoid with 0.1 A of current

    ANS: B　　　DIF: I　　　OBJ: 2-2.2

54. Why does a solenoid with an iron core produce a greater magnetic field than a solenoid without an iron core?
    a. The magnetic field produced by the solenoid causes the domains within the iron core to become better aligned.
    b. The magnetic field of the iron core adds to the magnetic field of the solenoid.
    c. Both the iron core and the solenoid contribute to the net magnetic field.
    d. all of the above

    ANS: D　　　DIF: I　　　OBJ: 2-2.2

55. When a current-carrying wire is placed between two poles of a magnet, the wire will
    a. move up or down, depending on the direction of the current in the wire.
    b. not move.
    c. circle clockwise or counterclockwise, depending on the direction of the current in the wire.
    d. None of the above

    ANS: A　　　DIF: II　　　OBJ: 2-2.2

56. In electric motors that use direct current, a device called a(n) ____ is attached to the armature to reverse the direction of the electric current in the wire.
    a. solenoid
    b. iron core
    c. commutator
    d. armature

    ANS: C　　　DIF: I　　　OBJ: 2-2.3

57. Galvanometers work by placing a(n) ____ between the poles of a permanent magnet.
    a. armature
    b. commutator
    c. electromagnet
    d. ferromagnet

    ANS: C   DIF: I   OBJ: 2-2.3

58. Electric motors work by placing a(n) ____ between the poles of a permanent magnet or an electromagnet.
    a. armature
    b. electromagnet
    c. galvanometer
    d. Both (a) and (b)

    ANS: A   DIF: I   OBJ: 2-2.3

59. If you move a magnet through a coil of wire, an electric current is induced. Which of the following would induce the largest current in the wire?
    a. moving the magnet slower
    b. moving the magnet faster
    c. reversing the direction of motion
    d. Both (b) and (c)

    ANS: B   DIF: I   OBJ: 2-3.1

60. What happens to the induced electric current in a solenoid if a magnet is pulled outward rather than pushed inward?
    a. It is no longer induced.
    b. It reverses direction.
    c. It decreases.
    d. It increases.

    ANS: B   DIF: I   OBJ: 2-3.1

61. What would induce a greater electric current in a wire?
    a. adding more loops of wire
    b. removing loops of wire
    c. pulling the magnet out rather than pushing it in
    d. pushing the magnet in rather than pulling it out

    ANS: A   DIF: I   OBJ: 2-3.1

62. Which of the following would not induce an electric current?
    a. moving a wire between the poles of a magnet
    b. moving a magnet in a coil of wire
    c. wrapping two wires around an iron ring and running a continuous current through one wire
    d. wrapping two wires around an iron ring and changing the current through one wire

    ANS: C   DIF: I   OBJ: 2-3.1

63. A device that uses electromagnetic induction to convert kinetic energy into electrical energy is called a(n)
    a. transformer.
    b. generator.
    c. commutator.
    d. armature.

    ANS: B   DIF: I   OBJ: 2-3.2

64. A device that uses electromagnetic induction to increase or decrease the voltage of an alternating current is called a(n)
   a. transformer.
   b. generator.
   c. commutator.
   d. armature.

   ANS: A   DIF: I   OBJ: 2-3.3

65. When the coil in a generator is not cutting through the magnetic field lines,
   a. the electric current increases.
   b. the electric current decreases.
   c. no electric current is induced.
   d. an electric current is induced.

   ANS: C   DIF: I   OBJ: 2-3.2

66. The electric current produced by a simple generator changes direction each time the coil makes a
   a. quarter-turn.
   b. half-turn.
   c. three-quarter-turn.
   d. whole turn.

   ANS: B   DIF: I   OBJ: 2-3.2

67. The primary coil of ____ has fewer loops than the secondary coil.
   a. a step-up transformer
   b. a step-down transformer
   c. an electromagnet
   d. a solenoid

   ANS: A   DIF: I   OBJ: 2-3.3

68. The secondary coil of a step-up transformer has
   a. more loops than the primary coil.
   b. fewer loops than the primary coil.
   c. the same number of loops as the primary coil.
   d. None of the above

   ANS: A   DIF: I   OBJ: 2-3.3

69. The secondary coil of a step-down transformer has
   a. more loops than the primary coil.
   b. fewer loops than the primary coil.
   c. the same number of loops as the primary coil.
   d. None of the above

   ANS: B   DIF: I   OBJ: 2-3.3

70. The primary coil of ____ has more loops than the secondary coil.
   a. a step-up transformer
   b. a step-down transformer
   c. an electromagnet
   d. a solenoid

   ANS: B   DIF: I   OBJ: 2-3.3

71. To function, transformers use
   a. an induced current.
   b. direct current.
   c. alternating current.
   d. a permanent magnet.

   ANS: C   DIF: I   OBJ: 2-3.3

Holt Science and Technology
Copyright © by Holt, Rinehart and Winston. All rights reserved.

72. Electric current carried by very high voltage transmission lines use a(n) ____ at a local distribution station before it goes to local power lines.
    a. electric generator
    b. electric motor
    c. step-up transformer
    d. step-down transformer

    ANS: D   DIF: I   OBJ: 2-3.3

73. Turbines generally turn the ____ in a generator, inducing an electric current.
    a. transformer
    b. magnet
    c. wire loop
    d. solenoid

    ANS: B   DIF: I   OBJ: 2-3.2

74. Which of the following is currently being used to turn turbines in generators?
    a. steam
    b. water
    c. wind
    d. All of the above

    ANS: D   DIF: I   OBJ: 2-3.2

75. The crystals of some kinds of molten rocks line up with the Earth's magnetic field as the rocks cool. However, in older layers of these rocks, the crystals line up in the direction opposite to the Earth's magnetic field. What does this suggest about the Earth's magnetic field?
    a. It never changes.
    b. It can reverse over long periods of time.
    c. It is too strong to be influenced by anything.
    d. None of the above

    ANS: B   DIF: III   OBJ: 2-1.4

76. Which of the following statements best describes what scientists think produces the Earth's magnetic field?
    a. The electric charges in the Earth's core do not move.
    b. The domains in the Earth's solid inner core are aligned.
    c. Electric charges move as liquid in the Earth's core flows.
    d. The Earth's core is a large bar magnet.

    ANS: A   DIF: III   OBJ: 2-1.4

COMPLETION

1. All magnets have two _____ where the magnetic effects are strongest. (poles or inductors)

    ANS: poles   DIF: I   OBJ: 2-1.1

2. Within an object are tiny magnetic _____ that can be either aligned or misaligned. (remains or domains)

    ANS: domains   DIF: I   OBJ: 2-1.2

3. _____ magnets retain their magnetic properties well but are difficult to magnetize. (Temporary or Permanent)

   ANS: Permanent   DIF: I   OBJ: 2-1.3

4. A(n) _____ converts kinetic energy into electrical energy. (electric motor or generator)

   ANS: generator   DIF: I   OBJ: 2-3.2

5. _____ occurs when an electric current is produced by a changing magnetic field. (Electromagnetic induction or Magnetic force)

   ANS: Electromagnetic induction   DIF: I   OBJ: 2-3.1

6. The interaction between electricity and magnetism is called _____. (electromagnetism or electromagnetic induction)

   ANS: electromagnetism   DIF: I   OBJ: 2-2.1

7. A _____ has two poles, exerts forces, and is surrounded by a magnetic field. (magnet or solenoid)

   ANS: magnet   DIF: I   OBJ: 2-1.1

8. An _____ changes electrical energy into kinetic energy. (electric motor or electromagnet)

   ANS: electric motor   DIF: I   OBJ: 2-2.3

9. Generators and transformers work on the principle of _____. (magnetic force or electromagnetic induction)

   ANS: electromagnetic induction   DIF: I   OBJ: 2-3.3

10. A(n) _____ is a coil of wire that produces a magnetic field when it is carrying an electric current. (solenoid or electric motor)

    ANS: solenoid   DIF: I   OBJ: 2-2.2

11. A compass may not always point north if you hold it near a(n) _____.

    ANS: electromagnet   DIF: II   OBJ: 2-2.1

12. A coil of wire that produces a magnetic field when carrying an electric current is a(n) _____.

    ANS: solenoid   DIF: I   OBJ: 2-2.2

13. Electric motors use _____ current.

    ANS: direct   DIF: II   OBJ: 2-2.3

14. A loop or coil of wire that rotates in an electric motor is a(n) _____.

    ANS: armature   DIF: I   OBJ: 2-2.3

15. A device used to measure current through the interaction of an electromagnet and a permanent magnet is called a(n) _____.

    ANS: galvanometer   DIF: I   OBJ: 2-2.3

16. _____ makes it possible to use transformers to bring electrical energy to your home.

    ANS: Electromagnetic induction   DIF: I   OBJ: 2-3.3

SHORT ANSWER

1. Name three properties of magnets.

   ANS:
   All magnets have two poles, exert magnetic forces, and are surrounded by a magnetic field.

   DIF: I   OBJ: 2-1.1

2. Why are some iron objects magnetic and others NOT magnetic?

   ANS:
   Iron objects are magnetic if most of their domains are aligned. If their domains are randomly arranged, they aren't magnetic.

   DIF: I   OBJ: 2-1.2

3. Suppose you have two bar magnets. One has its north and south poles marked, but the other one does not. Describe how you could use the first magnet to identify the poles of the second magnet.

   ANS:
   Hold one end of the unmarked magnet close to the marked north pole. If the two magnets attract each other, that end of the unmarked magnet is its south pole. If the magnets repel each other, that end of the unmarked magnet is its north pole.

   DIF: II   OBJ: 2-1.1

4. Name the metals used to make ferromagnets.

    ANS:
    Ferromagnets are made from iron, nickel, cobalt, or alloys of those metals.

    DIF: I   OBJ: 2-1.3

5. How are temporary magnets different from permanent magnets?

    ANS:
    Temporary magnets are easy to magnetize but lose their magnetization easily. *Permanent magnets* are difficult to magnetize but retain their magnetic properties for a long time.

    DIF: I   OBJ: 2-1.3

6. Why are auroras more commonly seen in places like Alaska and Australia than in places like Florida and Mexico?

    ANS:
    Auroras are most commonly seen near Earth's magnetic poles. Because Alaska and Australia are close to the Earth's magnetic poles, people living there are more likely to see auroras than people living in Florida and Mexico, which are far away from the Earth's magnetic poles.

    DIF: II   OBJ: 2-1.4

7. Describe what happens when you hold a compass close to a wire carrying a current.

    ANS:
    The needle may deflect and not point north.

    DIF: I   OBJ: 2-2.1

8. How is a solenoid like a bar magnet?

    ANS:
    A solenoid has a magnetic field similar to the magnetic field around a bar magnet.

    DIF: I   OBJ: 2-2.2

9. What makes the armature in an electric motor rotate?

    ANS:
    The motor's magnet exerts a force (up on one side, down on the other) on the armature, causing the armature to rotate.

    DIF: I   OBJ: 2-2.3

10. Explain how a solenoid works to make a doorbell ring.

    ANS:
    When the doorbell button is pushed, an electric current in the wire of the solenoid creates a magnetic field in and around the solenoid. The magnetic field pulls the iron rod through the solenoid and the rod strikes the bell.

    DIF: I   OBJ: 2-2.3

11. What do Hans Christian Oersted's experiments have to do with a galvanometer? Explain your answer.

    ANS:
    Oersted's work (that an electric current produces a magnetic field) led to electromagnets. A galvanometer measures the interaction between the magnetic field produced by an electromagnet and the magnetic field of a permanent magnet.

    DIF: II   OBJ: 2-2.3

12. How does a generator produce an electric current?

    ANS:
    A generator rotates a coil of wire through a magnetic field. The changing magnetic field around the wire induces an electric current in the wire.

    DIF: I   OBJ: 2-3.2

13. Explain why rotating either the coil or the magnet induces an electric current in a generator.

    ANS:
    You can rotate either the coil or the magnet in a generator because both processes will produce a changing magnetic field around the wire. The changing magnetic field induces an electric current in the wire.

    DIF: I   OBJ: 2-3.2

14. One reason why electric power plants do not distribute electrical energy as direct current is that direct current cannot be transformed. Explain why NOT.

    ANS:
    The electric current in the transformer's primary coil makes the iron ring an electromagnet. That electromagnet induces an electric current in the secondary coil only if the electromagnet's magnetic field changes. Direct current passing through the primary coil will not change the electromagnet's magnetic field, so no electric current will be produced in the secondary coil.

    DIF: III   OBJ: 2-3.3

15. Explain Oersted's 1820 discovery.

ANS:
Oersted discovered that electric current produces a magnetic field and that the direction of the magnetic field is dependent on the direction of the current.

DIF: I   OBJ: 2-2.1

16. Explain how an electromagnet works.

ANS:
An iron core has current-carrying wire wrapped around it. The current causes the iron core to become magnetic. The entire coil-wrapped mechanism becomes an electromagnet.

DIF: II   OBJ: 2-2.2

17. Name two ways a magnet can induce an electric current in a wire.

ANS:
A magnet can induce an electric current in a wire either by moving the wire in a magnet's magnetic field or by moving the magnet around the wire.

DIF: II   OBJ: 2-3.1

18. Describe the differences between a step-up transformer and a step-down transformer.

ANS:
A *step-up transformer* has fewer loops in its primary coil than in its secondary coil; it is used to increase the voltage of an electric current. A *step-down transformer* has more loops in its primary coil than in its secondary coil; it is used to decrease voltage.

DIF: I   OBJ: 2-3.3

19. Explain why auroras are seen mostly near the North and South Poles.

ANS:
Auroras are usually seen near Earth's magnetic poles. Earth's magnetic poles are located near Earth's geographic North and South Poles.

DIF: I   OBJ: 2-1.4

20. Compare the function of a generator with the function of an electric motor.

ANS:
The function of a generator is opposite of the function of an electric motor. A *generator* converts kinetic energy to electrical energy and an *electric motor* converts electrical energy to kinetic energy.

DIF: I   OBJ: 2-3.2

21. Explain why some pieces of iron are more magnetic than others?

    ANS:
    Some pieces of iron are more magnetic than others because the domains of the magnetic pieces are more aligned, while the domains of the less-magnetic pieces are randomly arranged.

    DIF: I   OBJ: 2-1.2

22. Use the following terms to create a concept map: *electromagnetism, electricity, magnetism, electromagnetic induction, generators, transformers*.

    ANS:
    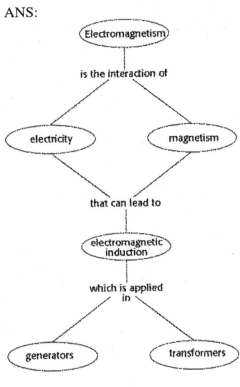

    DIF: II   OBJ: 2-3.2

23. You win a hand-powered flashlight as a prize in your school science fair. The flashlight has a clear plastic case so you can look inside to see how it works. When you press the handle, a gray ring spins between two coils of wire. The ends of the wire are connected to the light bulb. So when you press the handle, the light bulb glows. Explain how an electric current is produced to light the bulb. (Hint: Paper clips are attracted to the gray ring.)

    ANS:
    The electric current is produced by electromagnetic induction. The gray ring is a magnet, and when it spins, it creates a changing magnetic field around the coils of wire. An electric current is induced in the wire and is used to illuminate the light bulb.

    DIF: II   OBJ: 2-3.1

Holt Science and Technology

24. Fire doors are doors that can slow the spread of fire from room to room when they are closed. In some buildings, fire doors are held open by electromagnets. The electromagnets are controlled by the building's fire alarm system. If a fire is detected, the doors automatically shut. Explain why electromagnets are used instead of permanent magnets.

ANS:
Electromagnets are used instead of permanent magnets because electromagnets can be easily turned off. When the electromagnets are turned off, the fire doors are no longer held open, and they close.

DIF: II   OBJ: 2-2.3

25. Study the solenoids and electromagnets shown below. Rank them in order of strongest magnetic field to weakest magnetic field. Explain your ranking.

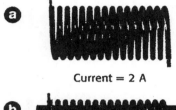
Current = 2 A

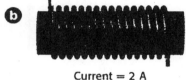
Current = 2 A

Current = 4 A

Current = 4 A

ANS:
Rank from strongest to weakest magnetic field: **(c), (b), (d), (a)**. Electromagnet **(c)** is strongest because it has an iron core and 4 A of current. Electromagnet **(b)** is next strongest because, even though is has a weaker current, it has an iron core. Solenoid **(d)** is next because it has a stronger current than **(a)**. Solenoid **(a)** is weakest.

DIF: II   OBJ: 2-2.2

26. Explain how a solenoid works.

    ANS:
    A solenoid is a coil of wire that produces a magnetic field when it is carrying an electric current. Because many loops of wire are used, the magnetic fields of the individual loops combine to produce a much stronger overall magnetic field.

    DIF: I   OBJ: 2-2.2

27.
    a. A current is produced when you move a magnet through a coil of wire. What factors increase the amount of electromagnetic induction?
    b. What factor changes the direction of the induced current?

    ANS:

    a. Electromagnetic induction is increased if the magnet is moved faster through the coil or if more loops are added to the coil surrounding the magnet.
    b. The direction of the induced current is reversed if the magnet is moved through the coil in the opposite direction.

    DIF: I   OBJ: 2-3.1

28. Monarch butterflies have magnetic material in their bodies. How might magnetic material be useful to them?

    ANS:
    Sample answer: Magnetic material could be useful when the Monarch butterflies migrate from one place to another. The magnetic material inside their bodies would align with the Earth's magnetic field and allow the butterflies to sense which direction is north as they fly.

    DIF: III   OBJ: 2-1.1

29. Use the following terms to complete the concept map below: *domains, atoms, copper, iron, north and south poles, electrons*.

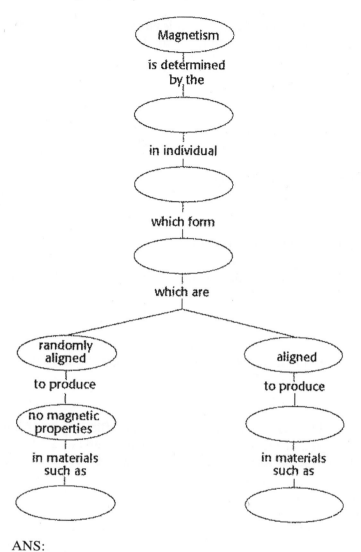

ANS:

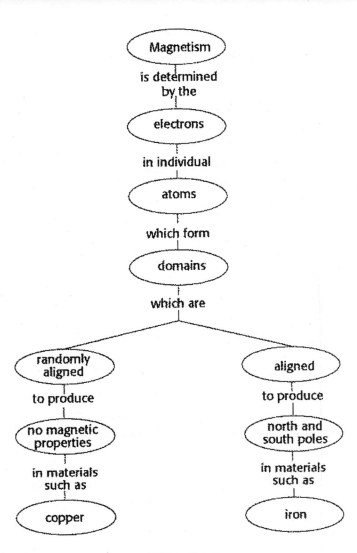

DIF: I   OBJ: 2-1.2

30. Examine the diagrams below, and answer the question that follows.

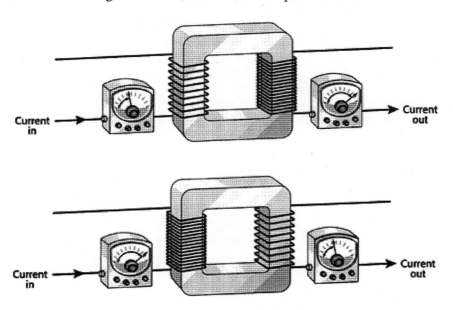

a. Which diagram above represents a step-up transformer? Explain your answer.
b. Which diagram above represents a step-down transformer? Explain your answer.

ANS:
a. The top diagram represents a step-up transformer because the primary coil has fewer loops than the secondary coil. Also, the voltage of the electric current leaving the transformer is higher than the voltage of the electric current entering it.
b. The bottom diagram represents a step-down transformer because the primary coil has more loops than the secondary coil. Also, the voltage of the electric current leaving the transformer is lower than the voltage of the electric current entering it.

DIF: I   OBJ: 2-3.3

31. How does a doorbell work?

ANS:
As the doorbell is pushed, it closes a circuit. The closed circuit allows an electric current to exist in the solenoid of the doorbell. The current produces a magnetic field. This field pulls an iron rod through the solenoid, and the rod strikes the bell.

DIF: II   OBJ: 2-2.3

32. How would you magnetize a screwdriver?

ANS:
Draw one end of a magnet in one direction along the metal part of the screwdriver.

DIF: II   OBJ: 2-1.2

Holt Science and Technology
Copyright © by Holt, Rinehart and Winston. All rights reserved.

ESSAY

1. Use your understanding of how a generator works to explain why people who drove the early automobiles had to turn a crank before they could start their car.

    ANS:
    An electric current was needed to start the car. People would induce this current by turning a crank that was connected to a wire loop. The loop was placed between the poles of a permanent magnet. As the rotating coil cut through the magnetic field lines of the magnet, an electric current was induced in the wire.

    DIF:　　III　　　　OBJ:　2-3.2

Short Course N Chapter 3—Electronic Technology

MULTIPLE CHOICE

1. All electronic devices transmit information using
    a. signals.
    b. electromagnetic waves.
    c. radio waves.
    d. modems.

    ANS: A   DIF: I   OBJ: 3-2.1

2. Semiconductors are used to make
    a. transistors.
    b. integrated circuits.
    c. diodes.
    d. All of the above

    ANS: D   DIF: I   OBJ: 3-1.1

3. Which of the following is an example of a telecommunication device?
    a. vacuum tube
    b. telephone
    c. radio
    d. Both (b) and (c)

    ANS: D   DIF: I   OBJ: 3-2.1

4. A monitor, printer, and speaker are examples of
    a. input devices.
    b. memory.
    c. computers.
    d. output devices.

    ANS: D   DIF: I   OBJ: 3-3.2

5. Record players play sounds that were recorded in the form of
    a. digital signals.
    b. electric current.
    c. analog signals.
    d. radio waves.

    ANS: C   DIF: I   OBJ: 3-2.3

6. Memory in a computer that is permanent and cannot be added to is called
    a. RAM.
    b. ROM.
    c. CPU.
    d. None of the above

    ANS: B   DIF: I   OBJ: 3-3.2

7. Cathode-ray tubes are used in
    a. telephones.
    b. telegraphs.
    c. televisions.
    d. radios.

    ANS: C   DIF: I   OBJ: 3-2.4

8. The central processing unit in a computer is NOT responsible for
    a. feeding data to the computer.
    b. doing calculations.
    c. solving problems.
    d. executing instructions.

    ANS: A   DIF: I   OBJ: 3-3.2

Holt Science and Technology
Copyright © by Holt, Rinehart and Winston. All rights reserved.

9. Which of the following is NOT true about semiconductors?
   a. They conduct electric current better than an insulator does.
   b. Silicon is the only element used in semiconductors.
   c. They do not conduct electric current as well as a conductor does.
   d. They are the backbone of most electronic devices.

   ANS: B DIF: I OBJ: 3-1.1

10. A vacuum tube will
    a. last longer than a transistor or a diode.
    b. take up less space than an integrated circuit.
    c. make telecommunication devices work faster.
    d. amplify electric current.

    ANS: D DIF: I OBJ: 3-1.4

11. A(n) ____ can convert alternating current to direct current.
    a. transistor
    b. diode
    c. signal
    d. amplifier

    ANS: B DIF: I OBJ: 3-1.2

12. An integrated circuit
    a. consists of one transistor and many silicon chips.
    b. operates at very slow speeds.
    c. has been a key factor in determining the size of electronic systems.
    d. requires electric charges to travel long distances.

    ANS: C DIF: I OBJ: 3-1.3

13. A collection of hundreds of tiny circuits that supply electric current to the various parts of an electronic device is called a
    a. circuit board.
    b. semiconductor.
    c. diode.
    d. transistor.

    ANS: A DIF: I OBJ: 3-1.1

14. A substance that conducts an electric current better than an insulator but not as well as a conductor is called a
    a. superconductor.
    b. semiconductor.
    c. diode.
    d. transistor.

    ANS: B DIF: I OBJ: 3-1.1

15. The process of replacing a few atoms of a semiconductor with a few atoms of another substance that have a different number of valence electrons is called
    a. integrating.
    b. segregating.
    c. doping.
    d. soldering.

    ANS: C DIF: I OBJ: 3-1.1

16. How can the conductivity of silicon be modified?
    a. through doping
    b. by replacing a silicon atom with an arsenic atom
    c. by replacing a silicon atom with a gallium atom
    d. All of the above

    ANS: D   DIF: I   OBJ: 3-1.1

17. Replacing a silicon atom with an arsenic atom in a silicon lattice results in
    a. an "extra" electron.   c. a "hole" where an electron could be.
    b. an "extra" proton.   d. a p-type semiconductor.

    ANS: A   DIF: I   OBJ: 3-1.1

18. Replacing a silicon atom with a gallium atom in a silicon lattice results in
    a. an "extra" electron.   c. an n-type semiconductor.
    b. an "extra" proton.   d. a p-type semiconductor.

    ANS: D   DIF: I   OBJ: 3-1.1

19. A free, unbonded electron quickly moves through a doped silicon lattice in
    a. a generator.   c. an n-type semiconductor.
    b. a transformer.   d. a p-type semiconductor.

    ANS: C   DIF: I   OBJ: 3-1.1

20. The way a semiconductor conducts electric current is based on
    a. how its electrons are arranged.   c. how its neutrons are arranged.
    b. how its protons are arranged.   d. All of the above

    ANS: A   DIF: I   OBJ: 3-1.1

21. An electronic component that allows electric current in only one direction is called a(n)
    a. integrated circuit.   c. diode.
    b. semiconductor.   d. transistor.

    ANS: C   DIF: I   OBJ: 3-1.2

22. An electronic component that can be used as an amplifier or as a switch is called a(n)
    a. integrated circuit.   c. diode.
    b. semiconductor.   d. transistor.

    ANS: D   DIF: I   OBJ: 3-1.2

23. An entire circuit containing many transistors and other electronic components formed on a single silicon chip is a(n)
    a. circuit board.   c. diode.
    b. integrated circuit.   d. transistor.

    ANS: B   DIF: I   OBJ: 3-1.2

24. What do you make when you join one layer of a p-type semiconductor with one layer of an n-type semiconductor?
    a. a diode
    b. a transistor
    c. an integrated circuit
    d. a circuit board

    ANS: A   DIF: II   OBJ: 3-1.2

25. What do you make when you sandwich one layer of a p-type semiconductor between two layers of an n-type semiconductor?
    a. a diode
    b. a PNP transistor
    c. an NPN transistor
    d. None of the above

    ANS: C   DIF: II   OBJ: 3-1.2

26. What do you make when you sandwich one layer of an n-type semiconductor between two layers of a p-type semiconductor?
    a. a diode
    b. a PNP transistor
    c. an NPN transistor
    d. None of the above

    ANS: B   DIF: II   OBJ: 3-1.2

27. Integrated circuits have caused
    a. electronic devices to increase in size.
    b. electronic systems to shrink.
    c. electrons in circuits to travel greater distances.
    d. the amount of time it takes for electrons to travel through a circuit to increase.

    ANS: B   DIF: I   OBJ: 3-1.3

28. Why can integrated circuit devices operate at high speeds?
    a. Electric charges must travel farther.
    b. Electric charges don't have as far to travel.
    c. Electric charges move faster.
    d. Electric charges become smaller and lighter.

    ANS: B   DIF: I   OBJ: 3-1.3

29. Why were early radios larger than modern radios?
    a. Early radios used vacuum tubes.
    b. Vacuum tubes in radios needed extra room to give off thermal energy.
    c. Modern radios use circuit boards.
    d. All of the above

    ANS: D   DIF: I   OBJ: 3-1.4

30. How are vacuum tubes similar to semiconductor diodes?
    a. Both can be used as a switch.
    b. Both can be used as an amplifier.
    c. Both can convert AC to DC.
    d. Both can convert DC to AC

    ANS: C   DIF: I   OBJ: 3-1.4

Holt Science and Technology
Copyright © by Holt, Rinehart and Winston. All rights reserved.

31. How are vacuum tubes similar to transistors?
    a. Both can be used as a switch.
    b. Both can be used as an amplifier.
    c. Both can convert AC to DC.
    d. both (a) and (b)

    ANS: D   DIF: I   OBJ: 3-1.4

32. Integrated circuits have enabled electronic devices
    a. to become much smaller.
    b. to perform more functions.
    c. to operate more quickly.
    d. All of the above

    ANS: D   DIF: I   OBJ: 3-1.3

33. Why aren't vacuum tubes used in modern electronics?
    a. They don't last as long as transistors and semiconductor diodes.
    b. They give off more thermal energy than transistors and semiconductor diodes.
    c. They are much larger than transistors and semiconductor diodes.
    d. All of the above

    ANS: D   DIF: I   OBJ: 3-1.4

34. Amplifiers increase the ____ of the current.
    a. speed
    b. wavelength
    c. amplitude
    d. frequency

    ANS: C   DIF: I   OBJ: 3-1.2

Each diagram below represents an electric current. Examine the diagrams and answer the questions that follow.

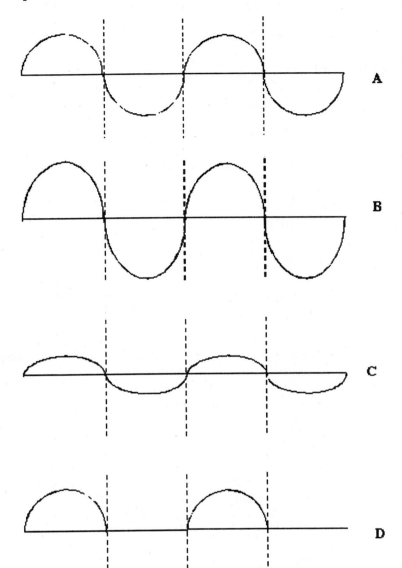

35. Electric current represented in diagram C can be converted to electric current represented in diagram B by using a(n)
    a. amplifier.
    b. switch.
    c. diode.
    d. AC adapter.

    ANS: A   DIF: II   OBJ: 3-1.2

36. Which diagram represents an electric current that would be the result of using an AC adapter?
    a. **A**
    b. **B**
    c. **C**
    d. **D**

    ANS: D   DIF: II   OBJ: 3-1.2

37. Which diagram represents alternating current?
    a. **A**
    b. **B**
    c. **C**
    d. All of the above

    ANS: D   DIF: II   OBJ: 3-1.2

38. A diode would allow the electric current represented by which diagram?
    a. **A**
    b. **B**
    c. **C**
    d. **D**

    ANS: D   DIF: II   OBJ: 3-1.2

39. Which diagram represents pulses of direct current?
    a. **A**
    b. **B**
    c. **C**
    d. **D**

    ANS: D   DIF: II   OBJ: 3-1.2

40. A signal whose properties, such as amplitude and frequency, can change continuously according to changes in the original information is called a(n)
    a. analog signal.
    b. digital signal.
    c. fiber optic signal.
    d. electronic signal.

    ANS: A   DIF: I   OBJ: 3-2.1

41. A series of electric pulses that represents the digits of binary numbers is called a(n)
    a. electronic signal.
    b. digital signal.
    c. fiber optic signal.
    d. analog signal.

    ANS: B   DIF: I   OBJ: 3-2.1

42. Which of the following is a part of a telephone?
    a. a transmitter
    b. a receiver
    c. wires
    d. All of the above

    ANS: D   DIF: I   OBJ: 3-2.2

43. When you speak into a telephone, a metal disk in the _____ vibrates due to sound waves.
    a. transmitter
    b. receiver
    c. wires
    d. phone line

    ANS: A   DIF: I   OBJ: 3-2.2

44. Vibrations in the transmitter of a telephone are converted into a(n)
    a. radio signal.
    b. digital signal.
    c. analog signal.
    d. All of the above

    ANS: C   DIF: III   OBJ: 3-2.2

45. An analog signal is converted back into a sound wave by a telephone
    a. transmitter.
    b. receiver.
    c. wire.
    d. amplifier.

    ANS: B   DIF: I   OBJ: 3-2.2

46. Long before record players there were Graphophones. Recordings were played on both Graphophones and record players by having a stylus ride in a continuous groove. Therefore, Graphophones played sounds that were recorded in the form of
    a. radio signals.
    b. digital signals.
    c. analog signals.
    d. All of the above

    ANS: C   DIF: II   OBJ: 3-2.3

47. Compact discs store a(n)
    a. analog signal.
    b. digital signal.
    c. radio signal.
    d. satellite signal.

    ANS: B   DIF: I   OBJ: 3-2.3

48. Each pit on a CD corresponds to the number
    a. 0.
    b. 1.
    c. 2.
    d. 3.

    ANS: A   DIF: I   OBJ: 3-2.3

49. Each nonpitted region on a CD, called a land, corresponds to the number
    a. 0.
    b. 1.
    c. 2.
    d. 3.

    ANS: B   DIF: I   OBJ: 3-2.3

50. A microphone creates an electric current that is a(n) ____ of the original sound wave.
    a. analog signal
    b. radio signal
    c. digital signal
    d. satellite signal

    ANS: A   DIF: I   OBJ: 3-2.4

51. A modulator combines the amplified analog signal from a microphone with
    a. radio waves of random frequencies.
    b. radio waves of a specific frequency.
    c. digital signals.
    d. video signals.

    ANS: B   DIF: I   OBJ: 3-2.4

52. A radio tower
    a. receives modulated radio waves through the air.
    b. transmits modulated radio waves through the air.
    c. combines modulated radio waves with other radio waves.
    d. generates noise.

    ANS: B   DIF: I   OBJ: 3-2.4

53. An antenna in a radio
    a. combines modulated radio waves with other radio waves.
    b. transmits modulated radio waves through the air.
    c. receives modulated radio waves through the air.
    d. generates noise.

    ANS: C   DIF: I   OBJ: 3-2.4

54. A radio's speakers convert a(n) ____ into sound.
    a. radio signal
    b. video signal
    c. analog signal
    d. digital signal

    ANS: C   DIF: I   OBJ: 3-2.4

55. An antenna in a television receives
    a. analog signals.
    b. digital signals.
    c. video signals
    d. radio signals.

    ANS: C   DIF: I   OBJ: 3-2.4

56. The images you see on your television are produced by
    a. beams of electrons projected onto a screen.
    b. laser signals projected onto a screen.
    c. radio signals projected onto a screen.
    d. analog signals projected onto a screen.

    ANS: A   DIF: I   OBJ: 3-2.4

57. An electronic device that performs tasks by processing and storing information is called a(n)
    a. microprocessor.
    b. computer.
    c. cellular device.
    d. integrated circuit.

    ANS: B   DIF: I   OBJ: 3-3.1

58. A computer performs a task
    a. when it is given a command.
    b. when it has all of the instructions it needs.
    c. when it is given input.
    d. All of the above

    ANS: D   DIF: I   OBJ: 3-3.1

Holt Science and Technology
Copyright © by Holt, Rinehart and Winston. All rights reserved.

Below is a flow chart illustrating the functions of a computer.
Examine the flow chart and answer the questions that follow.

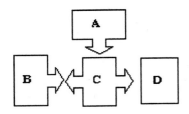

59. Which letter represents a computer's input?
    a. A   c. C
    b. B   d. D

    ANS: A   DIF: II   OBJ: 3-3.1

60. Which letter represents the output of a computer?
    a. A   c. C
    b. B   d. D

    ANS: D   DIF: II   OBJ: 3-3.1

61. Which letter represents the processing function of a computer?
    a. A   c. C
    b. B   d. D

    ANS: C   DIF: II   OBJ: 3-3.1

62. Which letter represents the storage function of a computer?
    a. A   c. C
    b. B   d. D

    ANS: B   DIF: II   OBJ: 3-3.1

63. The information you give to a computer is called
    a. input.   c. RAM.
    b. output.   d. ROM.

    ANS: A   DIF: I   OBJ: 3-3.1

64. When a computer _____ the input, it changes the input into a desirable form.
    a. stores   c. completes
    b. processes   d. All of the above

    ANS: B   DIF: I   OBJ: 3-3.1

65. What does a computer do if it doesn't process input right away?
    a. It stores the input.   c. It completes the input.
    b. It sends the input as output. d. All of the above

    ANS: A   DIF: I   OBJ: 3-3.1

Holt Science and Technology
Copyright © by Holt, Rinehart and Winston. All rights reserved.

66. Computers store information in their
    a. software.
    b. vacuum tubes.
    c. memory.
    d. diodes.

    ANS: C   DIF: I   OBJ: 3-3.1

67. The final result of a task performed by a computer is called
    a. input.
    b. output.
    c. throughput.
    d. memory.

    ANS: B   DIF: I   OBJ: 3-3.1

68. An integrated circuit that contains many of a computer's capabilities on a single silicon chip is a
    a. transistor.
    b. diode.
    c. microprocessor.
    d. circuit board.

    ANS: C   DIF: I   OBJ: 3-3.2

69. The parts or equipment that make up a computer is a computer's
    a. Internet connection.
    b. DSL line.
    c. hardware.
    d. software.

    ANS: C   DIF: I   OBJ: 3-3.2

70. A piece of hardware that feeds information to the computer is called a(n)
    a. input device.
    b. output device.
    c. throughput device.
    d. Internet connection.

    ANS: A   DIF: I   OBJ: 3-3.2

71. You can enter information into a computer using
    a. a keyboard and mouse.
    b. a scanner or digitizing pad and pen.
    c. a microphone.
    d. All of the above

    ANS: D   DIF: I   OBJ: 3-3.2

72. In a personal computer, the ____ is a microprocessor.
    a. RAM
    b. ROM
    c. CPU
    d. hard disk

    ANS: C   DIF: I   OBJ: 3-3.2

73. A computer does calculations, solves problems, and executes the instructions given to it in the
    a. RAM.
    b. ROM.
    c. CPU.
    d. hard disk.

    ANS: C   DIF: I   OBJ: 3-3.2

74. Which of the following would be LEAST likely to be found in a handheld, portable electronic game?
    a. memory
    b. ROM
    c. RAM
    d. modem

    ANS: D   DIF: II   OBJ: 3-3.2

75. Which of the following is permanent and handles functions such as computer start-up, maintenance, and hardware management?
    a. a floppy disk
    b. a CD-ROM
    c. RAM
    d. ROM

    ANS: D   DIF: I   OBJ: 3-3.2

76. Which of the following cannot be lost when the computer is turned off?
    a. RAM
    b. ROM
    c. an Internet connection
    d. input

    ANS: B   DIF: I   OBJ: 3-3.2

77. Working memory in which information is temporarily stored while that information is being used is called
    a. RAM.
    b. ROM.
    c. a CD-ROM.
    d. a hard disk.

    ANS: A   DIF: I   OBJ: 3-3.2

78. Increasing which of the following will make a more powerful computer by allowing more information to be input?
    a. a CD-ROM
    b. ROM
    c. RAM
    d. the CPU

    ANS: C   DIF: I   OBJ: 3-3.2

79. A portable zip drive is similar to a floppy drive, so it is a(n)
    a. input device.
    b. memory device.
    c. output device.
    d. throughput device.

    ANS: B   DIF: II   OBJ: 3-3.2

80. What piece of computer hardware serves as an input device as well as an output device?
    a. RAM
    b. ROM
    c. CPU
    d. modem

    ANS: D   DIF: I   OBJ: 3-3.2

81. A set of instructions, or commands, that tells a computer what to do is called
    a. hardware.
    b. software.
    c. memory.
    d. a microprocessor.

    ANS: B   DIF: I   OBJ: 3-3.3

82. Operating system software
    a. manages basic operations required by the computer.
    b. supervises all interactions between software and hardware.
    c. interprets commands from the input device.
    d. All of the above

    ANS: D   DIF: I   OBJ: 3-3.3

83. Application software contains instructions ordering the computer
    a. to operate a utility, such as a word processor.
    b. to access ROM.
    c. to store memory on the hard drive.
    d. All of the above

    ANS: A   DIF: I   OBJ: 3-3.3

84. Which of the following may connect to an Internet Service Provider?
    a. Local Area Network (LAN)   c. a business network
    b. a home computer   d. All of the above

    ANS: D   DIF: I   OBJ: 3-3.4

85. How do ISPs communicate with each other?
    a. using modems   c. via satellite
    b. through LANs   d. using cables

    ANS: C   DIF: I   OBJ: 3-3.4

COMPLETION

For each pair of terms, explain the difference in their meanings.

1. semiconductor/doping _____

   ANS: A *semiconductor* is a material that conducts electric current better than an insulator but not as well as a conductor. *Doping* a semiconductor changes its conductivity.

   DIF: I   OBJ: 3-1.1

2. transistor/diode _____

   ANS: A *transistor* is a device made of three layers of semiconductors that can act as an amplifier of a switch in a circuit. A *diode* is a device made of two layers of semiconductors that allows current in only one direction.

   DIF: I   OBJ: 3-1.2

3. signal/telecommunication _____

   ANS: A *signal* represents information. *Telecommunication* is the sending of information across long distances by electronic means.

   DIF: I   OBJ: 3-2.1

4. analog signal/digital signal _____

   ANS: An *analog signal* has properties that can change continuously with changes in the original information. A *digital signal* is a series of electric pulses that represent the digits of binary numbers.

   DIF: I   OBJ: 3-2.3

5. computer/microprocessor _____

   ANS: A *computer* is any electronic device that performs tasks by processing and storing information. A *microprocessor* is an integrated circuit containing many of a computer's capabilities on a single silicon chip.

   DIF: I   OBJ: 3-3.1

6. hardware/software _____

   ANS: *Hardware* refers to the parts that make up a computer. *Software* is the set of instructions that tells the computer what to do.

   DIF: I   OBJ: 3-3.2

7. A _____ can be used as an amplifier or as a switch. (transistor or semiconductor)

   ANS: transistor   DIF: I   OBJ: 3-1.2

8. A _____ allows electric current to travel in only one direction. (circuit board or diode)

   ANS: diode   DIF: I   OBJ: 3-1.2

9. Telegraphs were the first example of _____. (doping or telecommunication)

   ANS: telecommunication   DIF: I   OBJ: 3-2.1

10. The sounds you hear from a compact disc are a result of the sound being carried by a(n) _____. (analog signal or digital signal)

    ANS: digital signal   DIF: I   OBJ: 3-2.3

11. _____ is a set of instructions or commands that tell a computer what to do. (Software or Hardware)

    ANS: Hardware   DIF: I   OBJ: 3-3.2

12. Sending information across long distances electronically is called _____.

    ANS: telecommunication   DIF: I   OBJ: 3-2.1

13. Something that represents information, such as the international Morse code sent by a telegraph, is a _____.

    ANS: signal   DIF: I   OBJ: 3-2.1

14. Signals travel better when contained in another form of energy, called a _____.

    ANS: carrier   DIF: I   OBJ: 3-2.1

15. A wave that consists of changing electric and magnetic fields is an _____.

    ANS: electromagnetic wave   DIF: I   OBJ: 3-2.4

16. A huge computer network consisting of millions of computers that can all share information with one another is the _____.

    ANS: Internet   DIF: I   OBJ: 3-3.4

SHORT ANSWER

1. Describe how p-type and n-type semiconductors are made.

    ANS:
    N- and p-type semiconductors are both made by doping. An *n-type semiconductor* is made by replacing a semiconductor atom with an atom that has more than four electrons in its outermost energy level. A *p-type semiconductor* is made by replacing a semiconductor atom with an atom that has less than four electrons in its outermost energy level.

    DIF: I   OBJ: 3-1.1

2. Explain how a diode changes AC to DC.

    ANS:
    A diode allows current in only one direction. The direction of alternating current (AC) reverses, but the diode blocks the current in one direction. The result is pulses of direct current (DC).

    DIF: I   OBJ: 3-1.2

Holt Science and Technology
Copyright © by Holt, Rinehart and Winston. All rights reserved.

3. What two purposes do transistors serve?

   ANS:
   Transistors can be used to amplify an electric current or as a switch in a circuit.

   DIF: I   OBJ: 3-1.2

4. How might an electronic system that uses vacuum tubes be different from one that uses integrated circuits?

   ANS:
   An electronic system that uses a vacuum tube would be much larger than one that uses integrated circuits.

   DIF: II   OBJ: 3-1.3

5. How are analog signals different from digital signals?

   ANS:
   An *analog signal* has properties that can change with the original information. *Digital signals* are a series of electric pulses that represent binary numbers.

   DIF: I   OBJ: 3-2.3

6. Compare how a telephone and a radio tower transmit information.

   ANS:
   Both a telephone and a radio tower use analog signals to transmit information. However, telephone signals are sent over wires as electric current, while radio signals are sent out through the air as electromagnetic waves.

   DIF: I   OBJ: 3-2.4

7. How could a digital signal be corrupted?

   ANS:
   To corrupt a digital signal, information must be lost. A scratch on a CD, for example, could cause bits of information to be lost or misread, which could affect the sound.

   DIF: II   OBJ: 3-2.3

8. Using the terms input, output, processing, and memory, explain how you use a pocket calculator to add numbers.

    ANS:
    Sample answer: I use buttons on the calculator to *input* the numbers and the addition symbol into the calculator. The calculator accesses its *memory* to execute the addition command. *Processing* occurs when the calculator adds the numbers. The calculator shows the sum on its screen as *output*.

    DIF: I   OBJ: 3-3.2

9. What is the difference between hardware and software?

    ANS:
    Hardware is the parts and equipment that make up the computer.
    Software is a set of instructions that tells a computer what to do.

    DIF: I   OBJ: 3-3.3

10. Could something like the Internet exist without modems and telephone lines? Explain.

    ANS:
    Accept all reasonable answers. Sample answer: Yes; something like the Internet could exist using only satellites or radio to transmit signals from computer to computer. However, it would still require signals being sent, and something to receive the signals.

    DIF: II   OBJ: 3-3.4

11. What is the purpose of doping a semiconductor?

    ANS:
    A semiconductor is doped in order to alter its conductivity so that electric charges will flow.

    DIF: I   OBJ: 3-1.1

12. How does a transistor act as an amplifier?

    ANS:
    The small electric current in the transistor circuit can trigger a larger electric current in another circuit in the device.

    DIF: I   OBJ: 3-1.2

13. Describe an imaginary electronic device using the vocabulary words learned in this section.

    ANS:
    Answers will vary, but must reflect an understanding of the vocabulary words.

    DIF: I   OBJ: 3-2.4

14. Name three electronic telecommunication devices and the type of signal associated with each.

    ANS:
    Answers may vary. Sample answers: A telephone uses an analog signal. A compact disc uses a digital signal. A radio uses an analog signal.

    DIF: I   OBJ: 3-2.2

15. Compare the sound produced by analog and digital recording processes.

    ANS:
    Analog recordings are true to the original sound, but records wear out over time. *Digital recordings* can be played many times more than analog records. Noise can be easily removed from digital recordings.

    DIF: I   OBJ: 3-2.3

16. Discuss the changes in computers that have occurred since the development of ENIAC in 1946.

    ANS:
    Answers should include the changes in size brought about with the invention of the transistor and integrated circuits.

    DIF: I   OBJ: 3-3.1

17. Explain the difference between application software and operating system software.

    ANS:
    Application software involves a utility, like a computer game, word processor, or spreadsheet. *Operating system software* manages basic operations required by the computer and supervises software-to-hardware interactions.

    DIF: I   OBJ: 3-3.3

18. How is an electronic device different from a machine that uses electrical energy?

    ANS:
    An electronic device uses electrical energy to transmit information. A machine changes electrical energy into another form of energy that can be used to do work.

    DIF: I   OBJ: 3-1.3

19. How does a diode allow current to flow in one direction?

    ANS:
    When the terminals of a source of electrical energy are hooked up to a diode in the proper direction, charges can move between the semiconductor layers of the diode. If the terminals are reversed, the charges will move to opposite ends of the diode and no charges will move between the layers.

    DIF: I   OBJ: 3-1.2

20. In one or two sentences, describe how a TV works.

    ANS:
    A TV receives video signals that it separates into signals for three electron beams. The electron beams flow through the CRT, striking red, blue, and green fluorescent materials in the screen, forming the picture.

    DIF: I   OBJ: 3-2.4

21. Give three examples of how computers are used in your everyday life.

    ANS:
    Accept all reasonable responses. Sample answers: I use a digital alarm clock to wake up, I use a calculator in math class, and I use a computer to browse the World Wide Web.

    DIF: I   OBJ: 3-3.1

22. Explain the advantages that transistors have over vacuum tubes.

    ANS:
    Transistors are smaller, produce less thermal energy, and last longer than vacuum tubes.

    DIF: I   OBJ: 3-1.4

23. Use the following terms to create a concept map: *electronic devices, radio waves, electric current, signals, information.*

    ANS:

    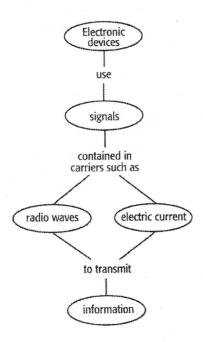

    DIF: II   OBJ: 3-2.4

24. Your friend is preparing an oral report on the history of radio and finds a photograph of a large, vintage radio. "Why is this radio so huge?" he asks you. Using what you know about electronic devices, how do you explain the size of this vintage radio?

    ANS:
    This radio was built before the invention of semiconductors. It was built with vacuum tubes instead of transistors. Vacuum tubes are much larger than transistors, so the radio is much larger than a modern radio.

    DIF: II   OBJ: 3-1.3

25. Using what you know about the differences between analog and digital signals, explain how the sound from a record player is different from the sound from a CD player.

    ANS:
    The sound produced by a record player comes from an *analog signal* of the sound stored on the record. The sound from a CD player comes from a *digital signal* of the sound stored on a CD. It is easier to remove noise from a recording on a CD.

    DIF: II   OBJ: 3-2.3

Holt Science and Technology
Copyright © by Holt, Rinehart and Winston. All rights reserved.

26. What do Morse code and digital signals have in common?

    ANS:
    Sample answer: Digital signals are represented by a combination of pulses and missing pulses of electric current. Likewise, Morse code signals are represented by combinations of dots and dashes.

    DIF: II   OBJ: 3-2.3

27. Computers can process a lot of information, but they cannot think. Explain why this is true.

    ANS:
    Computers can only process the information that is given to them. They can also only perform the functions for which they are programmed.

    DIF: III   OBJ: 3-3.1

28. Based on what you learned in the chapter, how do you think an automatic garage door opener works?

    ANS:
    Sample answer: The remote control sends a signal to the garage door opener. The signal activates a small motor, and the garage door opens.

    DIF: II   OBJ: 3-2.4

    Look at the diagram below, and answer the questions that follow.

    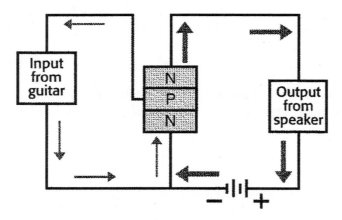

29. What purpose does the transistor serve in this situation?

    ANS:
    This transistor is an amplifier in the circuit.

    DIF: II   OBJ: 3-1.2

30. How does the current in the left side of the circuit compare with the current in the right side of the circuit?

    ANS:
    The current in the right side of the circuit varies in the same pattern as the current in the left side of the circuit; the current in the right side of the circuit is larger.

    DIF: II OBJ: 3-1.2

31. How does the sound from the speaker compare with the sound from the guitar?

    ANS:
    The sound from the speaker is the same as the sound from the guitar, except the sound from the speaker is louder.

    DIF: II OBJ: 3-1.2

32. Explain the difference between operating system software and application software.

    ANS:
    Operating system software manages basic operations required by the computer and supervises all interactions between software and hardware. It also interprets commands from input devices. *Application software* contains a set of instructions that order the computer to operate a utility.

    DIF: I OBJ: 3-3.3

33. Explain the difference between an analog recording and a digital recording.

    ANS:
    An *analog recording* reproduces sound waves that are similar to the original sound. For this reason, an analog recording also captures undesirable sounds. In a *digital recording*, the original sound waves are sampled many times so that undesirable sounds can be filtered out.

    DIF: I OBJ: 3-2.3

34. In 1946, the United States Army built ENIAC, the first general-purpose computer. ENIAC was so large that it could fill up an entire room. Which pieces of hardware do you think caused ENIAC to be so large? Explain.

    ANS:
    The central processing unit and the memory contain many electric circuits. When ENIAC was built, electric circuits contained massive vacuum tubes instead of transistors. These vacuum tubes, which were located in the central processing unit and the memory, caused ENIAC to be extremely large.

    DIF: II OBJ: 3-3.1

35. Use the following terms to complete the concept map below: *switches, diodes, transistors, one-way valves, electronic devices, semiconductors.*

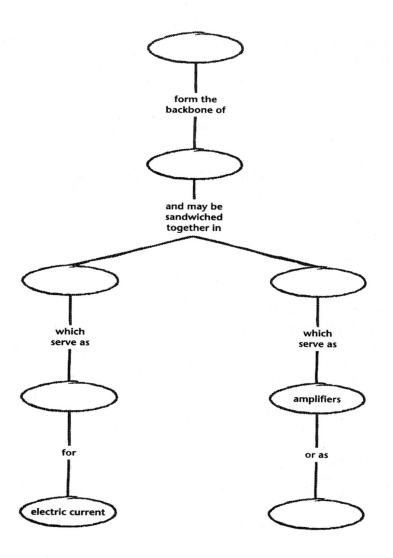

ANS:

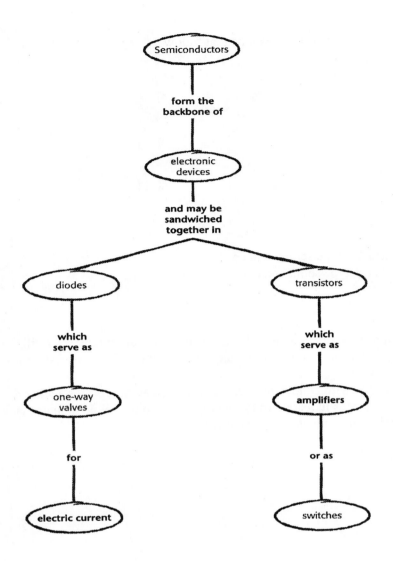

DIF: II OBJ: 3-1.2

36. Examine the diagram below, and answer the questions that follow.

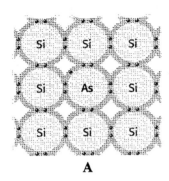 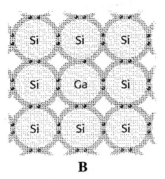

                A   B

a. Which diagram represents an n-type semiconductor? Explain.
b. Which diagram represents a p-type semiconductor? Explain.

ANS:

a. Diagram A represents an n-type semiconductor because an arsenic atom replaces a silicon atom. Arsenic has five electrons in its outermost energy level, giving the semiconductor an extra electron and a negative charge.
b. Diagram B represents a p-type semiconductor because a gallium atom replaces a silicon atom. Gallium has three electrons in its outermost energy level, so the semiconductor lacks an electron. This gives the semiconductor a positive charge.

DIF: II   OBJ: 3-1.1